GLOBALIZATION OF MATERIALS R&D

TIME FOR A NATIONAL STRATEGY

Committee on Globalization of Materials Research and Development

National Materials Advisory Board

Division on Engineering and Physical Sciences

NATIONAL RESEARCH COUNCIL
OF THE NATIONAL ACADEMIES

THE NATIONAL ACADEMIES PRESS
Washington, D.C.
www.nap.edu

THE NATIONAL ACADEMIES PRESS • 500 Fifth Street, N.W. • Washington, DC 20001

NOTICE: The project that is the subject of this report was approved by the Governing Board of the National Research Council, whose members are drawn from the councils of the National Academy of Sciences, the National Academy of Engineering, and the Institute of Medicine. The members of the committee responsible for the report were chosen for their special competences and with regard for appropriate balance.

This project was supported by the Department of Defense under Award No. MDA972-01-D-0001. Any opinions, findings, conclusions, or recommendations expressed in this publication are those of the author(s) and do not necessarily reflect the views of the sponsors.

International Standard Book Number 0-309-09603-0 (Book)
International Standard Book Number 0-309-54910-8 (PDF)
Library of Congress Control Number: 2005924163

Additional copies of this report are available from:

The National Academies Press, 500 Fifth Street, N.W., Washington, DC 20001; (800) 624-6242 or (202) 334-3313 (in the Washington metropolitan area); Internet <http://www.nap.edu>; and the National Materials Advisory Board, National Research Council, 500 Fifth Street, N.W., Washington, DC 20001; Internet <http://www.national-academies.org/nmab>.

Printed in the United States of America

THE NATIONAL ACADEMIES

Advisers to the Nation on Science, Engineering, and Medicine

The **National Academy of Sciences** is a private, nonprofit, self-perpetuating society of distinguished scholars engaged in scientific and engineering research, dedicated to the furtherance of science and technology and to their use for the general welfare. Upon the authority of the charter granted to it by the Congress in 1863, the Academy has a mandate that requires it to advise the federal government on scientific and technical matters. Dr. Ralph J. Cicerone is president of the National Academy of Sciences.

The **National Academy of Engineering** was established in 1964, under the charter of the National Academy of Sciences, as a parallel organization of outstanding engineers. It is autonomous in its administration and in the selection of its members, sharing with the National Academy of Sciences the responsibility for advising the federal government. The National Academy of Engineering also sponsors engineering programs aimed at meeting national needs, encourages education and research, and recognizes the superior achievements of engineers. Dr. Wm. A. Wulf is president of the National Academy of Engineering.

The **Institute of Medicine** was established in 1970 by the National Academy of Sciences to secure the services of eminent members of appropriate professions in the examination of policy matters pertaining to the health of the public. The Institute acts under the responsibility given to the National Academy of Sciences by its congressional charter to be an adviser to the federal government and, upon its own initiative, to identify issues of medical care, research, and education. Dr. Harvey V. Fineberg is president of the Institute of Medicine.

The **National Research Council** was organized by the National Academy of Sciences in 1916 to associate the broad community of science and technology with the Academy's purposes of furthering knowledge and advising the federal government. Functioning in accordance with general policies determined by the Academy, the Council has become the principal operating agency of both the National Academy of Sciences and the National Academy of Engineering in providing services to the government, the public, and the scientific and engineering communities. The Council is administered jointly by both Academies and the Institute of Medicine. Dr. Ralph J. Cicerone and Dr. Wm. A. Wulf are chair and vice chair, respectively, of the National Research Council.

www.national-academies.org

COMMITTEE ON GLOBALIZATION OF MATERIALS RESEARCH AND DEVELOPMENT

PETER BRIDENBAUGH, *Chair*
MILLER ADAMS, The Boeing Company
ASHISH ARORA, Carnegie Mellon University
GILBERT BENAVIDES, Sandia National Laboratories
UMA CHOWDHRY, DuPont Company
EDWARD DOWLING, DeBeers Group
GORDON GEIGER, University of Arizona
JENNIE HWANG, H-Technologies Group, Inc.
MICHAEL JAFFE, Rutgers University
ROBERT PFAHL, International Electronics Manufacturing Initiative
NATALIA TAMIRISA, International Monetary Fund
XISHAN XIE, University of Science and Technology Beijing

Staff

MICHAEL MOLONEY, Study Director (from September 2004)
TONI MARECHAUX, Study Director (December 2003–September 2004)
EMILY ANN MEYER, Research Associate (December 2003–April 2004)
MARTA VORNBROCK, Research Associate
LAURA TOTH, Senior Program Assistant

Preface

The Committee on Globalization of Materials Research and Development was appointed by the National Research Council (NRC) in December 2003 to assess the status and impacts of the globalization of materials science and engineering (MSE) research and development (MSE R&D). The panel was charged to do the following:

- Evaluate existing benchmarks as appropriate to assess the current situation and trends in materials research and development in the global community.
- Identify reasons why U.S. companies may or may not choose to depend on materials research carried out abroad. Assess current laws, policies, and regulations that affect these decisions.
- Identify advances in technology that are driving globalization of materials R&D.
- Assess the impact of the factors mentioned above on the U.S. economy and national security. Include the effect of foreign participation in domestic R&D and the effect of U.S. participation in foreign R&D.
- In light of the above, recommend actions to ensure U.S. access to current materials research and development.

The committee met four times during the course of the study to hear detailed presentations on the issues surrounding globalization and globalization's impact on the current state of MSE R&D, the U.S. economy, and national security. In addition, numerous private interviews were conducted with individuals and colleagues in academia, the federal research agencies, and industry. The committee also organized a poll of a self-selected sample of members of the materials community. The committee is grateful to several professional societies—the American Ceramic Society, the American Physical Society, the Federation of Materials Societies, the Materials Research Society, The Minerals, Metals and Materials Society, the Society for Biomaterials, and the Society of Manufacturing Engineers—for their assistance and to John Armor, Tia Benson-Tolle, Keith Bowman, James Daley, Duane B. Dimos, Robert Hawsey, Terry Lowe, John E. Marra, Ozden Ochoa, Greg Schoeppner, Robert Shull, and Kathleen Taylor for their valuable suggestions and their critical input to the committee's report.

Chapter 1 of this report defines MSE and globalization in the broadest sense and examines the history of globalization and R&D in general. Chapter 2 focuses on indicators for the emergence of global research activity in MSE. Chapter 3 updates the NRC report *Experiments in International Benchmarking of U.S. Research Fields* (2000) in some of the materials subfields. Chapter 4 examines various U.S. regulatory regimes—export, technology transfer, intellectual property, tax policy, immigration, environmental safety and health, and product approval—that might influence corporate R&D globalization decisions. Chapter 5 discusses the economic and national security impacts for the United States of the globalizing trends in MSE R&D. Chapter 6 presents a series of recommendations based on the conclusions drawn in each of the chapters and aimed at defining a strategy for maintaining access to critical, cutting-edge MSE R&D. Because this study was sponsored by the Department of Defense (DOD), the committee focused much of its attention on analyzing and recommending particular actions for DOD and its agencies.

It is clear to the committee that the United States and other leading industrial nations are experiencing the globalization of MSE R&D. While R&D is moving offshore to support manufacturing facilities in central Europe and Asia, a much more important aspect of globalization is the massive and accelerating investments that foreign governments, most notably China and India, are making in their own R&D infrastructures, particularly education. This trend is occurring at a time when such investments in the United States are falling. The enrollment of foreign students in graduate science and engineering education at U.S. universities is dropping rapidly and that of U.S. students is in free fall. The Organisation for Economic Co-operation and Development (OECD) reports that in 2000, the share of students in China graduating with engineering degrees was about 40 percent while for the United States it was about 5 percent. Clearly, the United States has a serious problem in education that must be addressed at the national/federal level if it is to maintain its leadership in innovation. The solution to this problem is too important to our future to be left to local decision makers. Like much of U.S. commerce, the U.S. defense and intelligence communities have been successful because they have had access to a one- or two-generation lead in critical technologies.

It is the committee's hope that the conclusions and recommendations in this report will help prepare the United States to deal effectively with the globalization of MSE R&D, secure the nation from future threats, and ensure continued access to the best domestic or foreign MSE R&D in the world.

Finally, I wish to thank all the committee members for their insights, inputs, and various contributions to this study. I also wish to thank the staff of the National Materials Advisory Board for their assistance in the development and execution of this study and in the production of this report.

Peter Bridenbaugh, *Chair*
Committee on Globalization of Materials Research and Development

Acknowledgment of Reviewers

This report has been reviewed in draft form by individuals chosen for their diverse perspectives and technical expertise, in accordance with procedures approved by the National Research Council's Report Review Committee. The purpose of this independent review is to provide candid and critical comments that will assist the institution in making its published report as sound as possible and to ensure that the report meets institutional standards for objectivity, evidence, and responsiveness to the study charge. The review comments and draft manuscript remain confidential to protect the integrity of the deliberative process. We wish to thank the following individuals for their review of this report:

John Allison, Ford Motor Company,
Siegfried S. Hecker, Los Alamos National Laboratory,
Don Hillebrand, Argonne National Laboratory,
Conilee G. Kirkpatrick, HRL Laboratories,
Neil E. Paton, private consultant,
Rustum Roy, Pennsylvania State University
Lyle H. Schwartz, private consultant
Richard S. Stein, University of Massachusetts,
Ellen D. Williams, University of Maryland,
Albert F. Yee, University of California, Irvine, and
Joel S. Yudken, AFL-CIO.

Although the reviewers listed above have provided many constructive comments and suggestions, they were not asked to endorse the conclusions or recommendations, nor did they see the final draft of the report before its release. The review of this report was overseen by Elsa Garmire, Dartmouth College. Appointed by the National Research Council, she was responsible for making certain that an independent examination of this report was carried out in accordance with institutional procedures and that all review comments were carefully considered. Responsibility for the final content of this report rests entirely with the authoring committee and the institution.

Contents

Executive Summary

Under the auspices of the National Research Council's National Materials Advisory Board and with the sponsorship of the Department of Defense, a study was conducted by the Committee on Globalization of Materials Research and Development to assess the current status of materials science and engineering research and development (MSE R&D) from a global perspective, to identify the drivers of U.S. companies' decisions to locate materials research in the United States or abroad, to assess the impact of the globalization of MSE R&D on the U.S. economy and national security, and to recommend actions to ensure continued U.S. access to critical MSE R&D. Globalization of MSE R&D is defined in this study as the worldwide expansion of MSE knowledge-creation centers as a result of U.S. and non-U.S. industry and government investments, along with increased worldwide collaboration, facilitated by information technology. Data from the National Science Foundation (NSF) and other sources[1] indicate increases in transnational academia-led R&D with international academic and industrial collaborators as well as in transnational corporation-led R&D with foreign affiliates of U.S. corporations, foreign academics, or foreign corporations. Given the prospect that global shifts in MSE R&D will continue, What are the possible risks and benefits for the

[1]See NSF, *Main Science and Technology Indicators*, 2004, available at http://www.nsf.gov/sbe/srs/seind04/start.htm; Economist Intelligence Unit, *Scattering the Seeds of Invention: The Globalization of Research and Development*, 2004; and the results of a poll of MSE practitioners carried out for this study as described in Appendix E.

United States? Like the study of materials itself, the task of assessing current and future impacts of the globalization of MSE R&D is complex and involves interconnected and multidimensional elements.

Today global MSE R&D is diversifying geographically at an accelerating rate as various other countries, including some not previously known as centers of MSE expertise, invest in the creation of their own MSE knowledge base. As a result, the relative U.S. position in many MSE subfields is in a state of flux. The European Union and the Asia-Pacific region, most notably Japan and most recently China, are now challenging traditional U.S. leadership in various subfields of MSE; it appears, for instance, that Japan has surpassed the United States in the area of alloys and will surpass it in ceramics. In all the MSE subfields examined during this study, global R&D is diversifying and becoming increasingly dispersed geographically, but how this trend will evolve and what the full impact will be for the United States are not yet clear.

CONCLUSIONS ON THE CURRENT SITUATION

The committee offers these conclusions on MSE R&D today:

Conclusion. Globalization of MSE R&D is proceeding rapidly, in line with broader trends toward globalization. As a result of increasing international trade and investment, the emergence of new markets, and the growth of the Internet and the global communications system, MSE R&D in the United States is an internationalized activity with a diverse set of international partners.

Conclusion. The globalization of MSE R&D is narrowing the technological lead of the United States.

Patent and literature surveys suggest that at the moment the United States remains either the world leader or among the world leaders across the MSE subfields. The benchmarking evidence in this report and from a previous study[2] paints a varied picture across the MSE subfields, indicating that the United States leads in some critical areas and is among the leaders in others. In some subfields, however, all the data suggest that the probability of the United States' maintaining leadership in MSE R&D varies from uncertain to unlikely.

[2]National Academy of Science, National Academy of Engineering, Institute of Medicine, *Experiments in International Benchmarking of U.S. Research Fields*, Committee on Science and Engineering Public Policy, Washington, D.C.: National Academy Press (2000).

Conclusion. At this stage, economic analysis is limited by a dearth of data and by the lack of a comprehensive empirical framework. Although available evidence suggests that the globalization of MSE R&D has had a limited impact on the U.S. economy so far, the medium-term impact is highly uncertain. A positive impact will depend on globalized MSE R&D leading to increased U.S. productivity and contributing positively to U.S. domestic innovation.

The impact on the U.S. economy of globalized MSE R&D is likely to differ across materials subfields. On the one hand, a decline in domestic MSE R&D in particular subfields might have a negative effect on domestic growth, wages, and jobs in those and other MSE R&D subfields and industries dependent on materials research. On the other hand, a relative decline in MSE R&D in one subfield might release resources for investment in another, more promising subfield in which the United States enjoys a comparative advantage, thus enabling U.S. firms to generate new knowledge, products, and growth in the medium term. Similarly, relocating overseas any MSE R&D that can be performed more efficiently by foreign counterparts might allow U.S. firms to expand other domestic MSE R&D, thereby increasing the global knowledge base that will stimulate innovation in all countries. One result could be a new comparative advantage for the United States if it can integrate the results of domestic and global research to create new, higher-value products. On balance, the United States may well gain from globalization of MSE R&D, provided that conditions in the private and public sectors lead to increased U.S. productivity, efficiency, and capacity for innovation.

Conclusion. The results of MSE R&D continue to enhance U.S. national security and homeland defense by adding improved materials capabilities to the weapons and protective systems used by today's warfighters. The evolution of materials research in the United States and abroad will affect the nation's ability not only to defend against emerging threats of the 21st century but also to ensure a healthy economy as a basic underpinning of national security. Because knowledge and the intellectual capacity to generate new knowledge are proliferating across the world, because innovation and development cycles are becoming shorter, and because U.S. dependence on foreign sources of innovation is increasing, the lead in critical technologies enjoyed thus far by the U.S. defense and intelligence communities will be seriously eroded without mitigating action.

The ability to meet 21st century U.S. defense needs will depend on R&D in materials and processes to improve existing materials and achieve breakthroughs in new materials and combinations, including, for example, lightweight materials that provide equivalent functionality, materials that enhance protection and survivability, and materials that improve propulsion technology. Future defense systems could employ advanced materials that are self-healing, that interact independently with the local environment, that monitor the health of a structure or component during operation, or that host evolving technologies, such as embedded sensors and integrated antennas. Such materials must also deliver traditional high performance in structures; protect against corrosion, fouling, erosion, and fire; control fractures; and serve as fuels, lubricants, and hydraulic fluids. Requirements for material producibility, low cost, and ready availability will be much more demanding than they are today.

Conclusion. In response to the globalization of MSE R&D, it is the task of public policy to minimize the risks and maximize the benefits to ensure the ongoing U.S. innovation that is essential to the nation's economy and national security and to facilitate continued access to the new knowledge generated by MSE R&D.

The impact of the globalization of MSE R&D can be positive and large, but the risks of a negative impact for the United States remain substantial. Available data show that companies globalize their R&D for a number of reasons, including the availability of expertise, the impacts of regulatory regimes, proximity to new international customers, and cost savings. Risk factors for U.S. corporate investment in R&D overseas are varied but can include concern about the ownership of intellectual property and the security of trade secrets, as well as wider concerns about the rule of law and democratic institutions, particularly in developing economies. To ensure a positive impact, the U.S. government and the private sector must exploit foreign or joint R&D to benefit domestic innovation by integrating it efficiently and effectively into domestic R&D programs, both civilian and military.

With the emergence of new centers of high-value research across the globe has come a new, marketlike demand for the world's finest students and experts, challenging the ability of the United States to attract top researchers. Any reduction in the supply of non-U.S. experts involved directly in U.S. research and innovation, along with the acknowledged difficulty of attracting U.S. citizens to MSE, will constrain the supply of top scientists and engineers within the United States ready to conduct the MSE R&D needed for U.S. economic growth and national security. Any such loss in expertise will diminish not only the value of the U.S. research output but also, in the long term, the nation's capacity to recognize, understand, and exploit the research output of the rest of the world.

Even if the United States makes great efforts to maintain control of U.S.-generated technologies, knowledge, and capabilities, other governments' investments in their own MSE R&D will challenge the ability of the United States to lead technologically. The loss of a U.S. national capacity for MSE research, or a decline in the ability of U.S. manufacturing to take advantage of and also to motivate MSE research, or a diminished U.S. military, homeland defense, or intelligence capability is not merely a matter of national pride or international image. In a knowledge-based future, only if the United States continues to have access to, and in many cases generate, cutting-edge science and technology will it be able to sustain its current world economic leadership and its strength in national defense and security.

Conclusion. It is in the long-term interest of the United States to participate in international partnerships in MSE R&D and thereby ensure U.S. access to cutting-edge knowledge and technology.

Conclusion. There is a need to maintain a robust U.S. MSE R&D infrastructure whereby materials problems can be addressed and solved and the solutions verified, from laboratory through pilot scale.

Conclusion. The MSE education system, including K–12 mathematics and science education, will have to evolve and adapt so as to ensure a supply of MSE professionals educated to meet U.S. national needs for MSE expertise and to compete on the global MSE R&D stage. The evolution of the U.S. education system will have to take into account the materials needs identified by the federal agencies that support MSE R&D as well the needs of the materials industry.

RECOMMENDATIONS FOR ACTION

How can the United States best maintain access to the global output of critical MSE R&D? How can it sustain a leadership position in the creation and use of new MSE knowledge? How can it ensure full access to knowledge and technology or access to the right sort of knowledge and technology? Integration also must be a priority, but integrating R&D is not easy. There is a risk that some knowledge generated in MSE R&D abroad will not be absorbed in the United States and that domestic U.S. expertise may not be sufficient to recognize foreign innovation and maximize its integration. Maintaining access to current MSE R&D will require active management, which in turn requires that the nation's public policy and government leaders ensure development of a national strategy that allows the United States to benefit from all that the globalization of MSE R&D has to offer.

RECOMMENDATION ON DEVELOPING A NATIONAL STRATEGY

To maximize the benefits for the United States of the globalization of materials science and engineering research and development (MSE R&D), the federal government should create a well-defined and coordinated national strategy to manage the development of and access to strategic MSE knowledge and technology in a global framework. Particular emphasis must be given to defining and achieving MSE R&D goals for ensuring a strong 21st century U.S. military and a secure U.S. homeland.

In building a U.S. national strategy for effective development and use of MSE R&D, the following elements should be considered:

- Identifying in MSE R&D across the defense services and other relevant national security agencies programmatic linkages that will facilitate a coordinated approach to answering critical questions across the subfields of MSE and assessing the readiness of R&D programs to do so, analyzing domestic readiness to provide critical MSE capabilities, and developing recommendations on the role that international and transnational MSE R&D might play;
- Defining (1) immediate priorities for which programmatic directions are clear and (2) next steps, which will require development of a roadmap as a prelude to determining relevant MSE R&D programs;
- Including as participants a comprehensive range of stakeholders and decision makers from the defense, homeland security, and intelligence communities and obtaining significant input from and coordinating with the wider federal science and engineering agencies—including the National Science Foundation, the Department of Energy, NASA, and so on; and
- Soliciting independent advice from academia, industry, and other experts, as required—perhaps with the participation of the Defense Science Board—and obtaining input from industry regarding policies and incentives that could encourage proactive industry strategies for sustaining a strong MSE R&D base in the United States.

The committee recognizes that building a robust and effective national strategy for ensuring U.S. access to the results of MSE R&D will also require a better understanding of current trends in MSE R&D worldwide; a clear and focused set of critical questions and challenges MSE R&D must address to meet national economic, defense, and homeland security needs; and a fresh approach to managing regulatory regimes, improving the education system, and strengthening the infrastructure for U.S. MSE R&D.

Any national strategy for ensuring U.S. access to ongoing MSE R&D will require not only sufficient information on global MSE R&D activity and better monitoring of it but also regularly updated benchmarking of the relative global status of U.S. MSE R&D. The committee quickly became aware of the lack of current data on the global flow of investments in R&D generally and in MSE R&D specifically. Building a national strategy for ensuring U.S. access to MSE R&D will require better data and new analytical tools to deal with the complexity of the R&D globalization phenomenon, and obtaining these data and tools will require the collective effort of various agencies across the federal government. Developing the tools to maintain continuous access to global R&D will require consideration of how Department of Defense (DOD) technology forecasting and monitoring systems, for instance, can be strengthened and how current DOD initiatives aimed at identifying critical technology worldwide can be expanded.

Essential for a successful strategy is a thorough understanding of what knowledge is needed to develop effective national defense and homeland security systems and how and from where it can be obtained. What critical capabilities will a strong 21st century U.S. defense capability require? Of the priorities already suggested[3] or still to be determined, how will choices be made and MSE R&D used to help achieve necessary improvements and breakthroughs in materials? The committee emphasizes that neither innovation nor future threats can always be foreseen and predicted—it may not always be clear today what new capabilities might be developed from the results of yesterday's research or what particular challenges tomorrow's adversaries might present. Moreover, previous success in acquiring technology from other countries does not guarantee that it can be acquired from them in the future. Addressing wider national security concerns effectively will benefit from the highest level of coordination and cooperation within DOD and between relevant federal agencies; from ongoing assessment of existing critical technology lists, contractual arrangements, and R&D funding procedures; and from the definition of longer-term goals and challenges for MSE R&D.

In the rapidly evolving environment for MSE R&D, how should the nation's regulatory regime take into account the realities of R&D in the 21st century? Regulatory regimes can drive decisions on where R&D is performed. Among the many considerations are concerns about the security of intellectual property developed abroad, the effects of the export licensing process on the execution of

[3]See the following reports: Defense Science Board, *Defense Science and Technology*, available at http://www.acq.osd.mil/dsb/reports/sandt.pdf; National Research Council, *Making the Nation Safer: The Role of Science and Technology in Countering Terrorism*, Washington, D.C.: The National Academies Press (2002); and National Research Council, *Materials Research to Meet 21st Century Defense Needs*, Washington, D.C.: The National Academies Press (2003).

R&D programs, the availability of skilled researchers, and provisions for tax incentives in the locales under consideration. As MSE R&D becomes increasingly global, it is important for public-policy makers to ensure that U.S. regulatory regimes do not unreasonably impede U.S. researchers' participation in international R&D of national importance or foreign researchers' participation in U.S. research. In addition, a review of the nation's regulatory system should ask whether there are technologies to which the nation must secure access but that it need not necessarily control, as well as how such a distinction would affect the nation's export control regime.

When building a national strategy it may be tempting to consider protecting U.S. interests by retreating from the world stage in areas deemed critical to U.S. national security or economic interests. A protectionist approach, however, might result in the United States' not having access to superior technologies developed elsewhere. Access to cutting-edge knowledge and technology can be better and more effectively achieved, the committee believes, if the United States becomes the most active player in global MSE R&D.

Maintaining a strong domestic U.S. capability to engage in international MSE R&D, to integrate non-U.S. MSE R&D into U.S. systems and ongoing U.S. R&D, and to monitor and understand global MSE R&D and its impact on U.S. leadership and capabilities will require that the U.S. educational system adapt accordingly, producing MSE practitioners who can achieve these goals. A very important, perhaps less-well-known aspect of globalization is the massive and accelerating investment being made by foreign governments, most notably China and India, in their R&D infrastructure, particularly education. These investments are occurring at the same time as comparable investments in the United States are falling. The Organisation for Economic Co-operation and Development (OECD) reports that in 2000 about 40 percent of students in China graduated with engineering degrees,[4] whereas in the United States the figure was about 5 percent. Clearly, this and similar trends in U.S. education must be addressed if the United States is to maintain its leadership in innovation, and a robust national research infrastructure must be in place.

Building a national strategy to ensure U.S. leadership in and access to advances in globalized MSE R&D will require specific efforts, and in this connection, the committee offers five more recommendations:

---

[4]OECD, *Education Statistics and Indicators, Education at a Glance 2002*, available at http://www.oecd.org/education.

RECOMMENDATION ON GATHERING BETTER DATA

U.S. data collection efforts and forecasting systems should be strengthened in order to monitor trends in the offshoring of MSE R&D and the growth of MSE R&D worldwide.

RECOMMENDATION ON IMPROVING MONITORING

The Department of Defense should build on existing capacities to monitor, assess, and promote access to developments in MSE R&D across the globe with a strategic view to underpinning the maintenance of U.S. leadership and security. In addition, existing U.S. government internal systems for strategic and critical technology analysis, management, and integration should be strengthened. Modern database and communication systems for identifying synergies across the defense services should be developed.

RECOMMENDATION ON CONDUCTING COMPREHENSIVE, EXPEDITED BENCHMARKING

An expedited benchmarking study, similar to Experiments in International Benchmarking of U.S. Research Fields *(National Academy Press, Washington, D.C., 2000), should be conducted immediately to assess the relative global position of the United States in MSE R&D.*

RECOMMENDATION ON ESTABLISHING LONG-TERM SECURITY NEEDS AND CHALLENGES

The Department of Defense should strengthen current systems for establishing clearly the materials needs of the 21st century warfighter as well as those essential to achieving national and homeland security priorities. Efforts in this regard should focus not on meeting the shorter-term acquisition needs of the military, but rather on identifying and prioritizing the longer-term questions and challenges that MSE R&D will have to address in order to meet identified long-term U.S. security needs.

RECOMMENDATION ON REVIEWING REGULATORY REGIMES

A systematic review of the rationale for and the impacts of U.S. government regulation of the transfer of knowledge and innovation across borders within the framework of globalized MSE R&D should be carried out by a government task force of representatives from the relevant agencies, with input from academia and industry.

In summary, the challenge presented by the globalization of MSE R&D is significant, multidimensional, and intrinsically interconnected across many agencies within the federal government. A national strategy to ensure U.S. leadership in and access to advances in global MSE R&D should be established and implemented as a national priority. Such a strategy should address the needs discussed in this study and should mitigate the risks identified. This report's recommendations offer a framework for the development of a robust strategy aimed at ensuring a positive impact for the United States and continued access in the future to cutting-edge MSE R&D worldwide.

1

Materials as Global Activity: Setting the Scene

Materials science and engineering (MSE) involves the generation and application of knowledge relating the composition, structure, and processing of materials to their properties and uses (see Box 1.1, "What Is Materials Science and Engineering?"). The science aspect focuses on discovering the nature of materials, which in turn leads to theories or descriptions that explain how structure relates to composition, properties, and behavior. The engineering aspect deals with the use of science to develop, prepare, modify, and apply materials to meet specific needs. The connection between the research and development (R&D) and science and engineering can be thought of in the following way: Research is an enabling process for science, and development is an enabling process for engineering.

At the beginning of the 21st century, MSE innovation and progress remain as dynamic as at any time in the history of the field. Over the last half of the 20th century, MSE remained a driver of economic activity in the United States and around the world. It can be thought of as the key building block of most advanced technologies, if not all, and as such it continues to make innumerable contributions to social advancement, human health and development, and the maintenance of national security. MSE contributions have included enhanced global communications systems that enable 21st century connectivity; materials relevant to and/or inspired by biology; and new materials for augmented security and

BOX 1.1 What Is Materials Science and Engineering?

Where Did MSE Come From?

Although materials and processes have fueled technological progress for thousands of years, the field of MSE per se did not exist before the 1960s. It is, therefore, a relatively young discipline in comparison with physics, chemistry, and related engineering fields. MSE became a single discipline through the evolution and coalescence of three materials-specific fields—metallurgy, ceramics, and polymer science. Although many other disciplines—for example, physics, geology, electronics, optics, chemistry, and biology—continue to bear on MSE and have made indispensable contributions to its development as a formal discipline, these three materials-based fields remain at the heart of MSE.

In the early days of MSE as an academic discipline and a subject of R&D endeavor, practitioners came mostly from physics, chemistry, engineering, metallurgy, the earth sciences, and mathematics. With the growth of biomaterials, medical practitioners, biologists, biochemists, and biophysicists have joined in. Increasingly, as the field emerged in its own right, MSE practitioners were trained in materials departments established at engineering or physical sciences schools in universities here and abroad. Nevertheless, despite 50 years of developing, maturing, and gaining broad acceptance, agreeing on an all-encompassing definition for MSE as a discipline remains a challenge. The origins and nature of MSE remain varied and interwoven, and any definition of the field must reflect the richness and diversity of all the activity related to "materials."[a]

What Is a Material?

A good place to start defining MSE is to consider what a material is. A simple definition would be that a material is the stuff from which an article, fabric, or structure is made.[b] This definition, however attractive because of its simplicity, does not reflect the full diversity of the study of materials. Because most articles, fabrics, or structures are considered to be solids, how would the study of liquids and gases fit into such a definition of materials? It would not, yet the study of liquids and gases is of central importance to many areas of MSE, such as materials processing, understanding the structure of many biological systems, investigating colloidal systems, and studying liquid crystals. A more thorough definition might be this: Matter is a "material" when that form of matter has structural, optical, magnetic, or electrical use.[c]

---

[a]NRC, *Materials and Man's Needs: Materials Science and Engineering*, Washington, D.C.: National Academy Press (1979); *Materials Science and Engineering for the 1990s: Maintaining Competitiveness in the Age of Materials*, Washington, D.C.: National Academy Press (1989).

[b]Definition based on that found in the Oxford English Dictionary.

[c]From *Aims and Objectives of a Degree in Materials Science*, written by Adrian Sutton, Department of Materials, Oxford University, United Kingdom.

What Do Materials Scientists and Engineers Do?

Further insight into what is meant by MSE can be gained by considering the kinds of things materials scientists study and the materials-related knowledge and skills they need to do so.[d]

- *Structure.* Electronic, atomic, bonding, crystalline, amorphous, and multiphase structure on the nano-, micro-, meso-, and macroscales.
- *Characterization of composition and microstructure.* Spectroscopy, optical and electron microscopy, electron and X-ray diffraction, scanning probe techniques, thermal analysis, and some aspects of traditional chemical analysis.
- *Phase equilibria and phase transformations.* Thermodynamic and kinetic aspects.
- *Mechanical behavior.* Elastic and plastic deformation and fracture; strengthening, toughening, and stiffening mechanisms; mechanical test methods; and continuum mechanics.
- *Functional behavior.* Semiconducting, dielectric, optical, conducting, and magnetic materials and materials that interact with or draw inspiration from biological systems.
- *Processing and manufacture.* Processing and synthesis of materials via gaseous, liquid, colloidal, powder, solid state, and deposition techniques; joining and fabrication methods; surface treatment; heat and mass transfer; and fluid mechanics.
- *Degradation and durability of materials.* Effect of liquid and gaseous environments on the performance of different material types, wear of material, and biodegradation.
- *Materials selection.* Consideration of all material types, including material processing method, life-cycle analysis and product costs; selection criteria for materials and production processes.
- *Design with materials.* The selection of appropriate compositions; choice of and use of processing and manufacture to achieve the required microstructure, structural, and functional properties in a product according to agreed specifications.

It has been suggested that these activities can be summarized by considering that materials scientists and engineers investigate the function of a material in an existing application and discover applications for it through the characterization of its structure and properties, and through the understanding of its production. The materials tetrahedron (Figure 1.1.1) is often used to illustrate the four aspects of MSE: (1) composition and microstructure, (2) properties, (3) synthesis and processing, and (4) performance. Each is

continues

[d]U.K. Quality Assurance Agency for Higher Education, *Materials* (2002). Available at http://www.qaa.ac.uk.

BOX 1.1 Continued

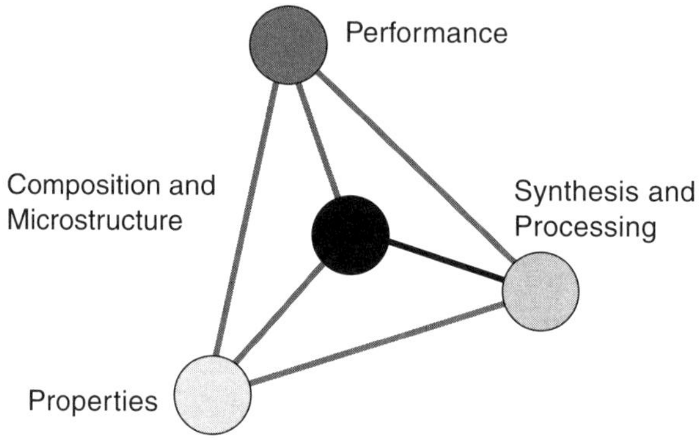

FIGURE 1.1.1 The MSE tetrahedron reflects the four fundamental aspects of a material—performance, properties, synthesis and processing, and composition and microstructure. While the four aspects can be treated individually as disciplines, a comprehensive approach to the materials tetrahedron is usually required for complex problems.

intellectually rich and challenging, and together they help define the discipline of MSE in its own right.

Because of the enormous breadth of the field, MSE R&D must be divided into subfields. In its 1993 and 1995 reports, the NSTC listed the subfields of materials research and engineering as biomaterials, ceramics, composites, electronic materials, magnetic materials, metals, optical-photonic materials, polymers, and superconducting materials. The com-

defense systems to protect the nation's security.[1] The quest for greater insight and innovation remains the primary motivation for the continuing evolution of a field that is characteristically inter- and multidisciplinary.

The evolution of MSE has been characterized not only by a broadening of the field to encompass new materials, such as nanomaterials, and new crosscutting themes, such as biomaterials, but also by the globalization of MSE industrial activity and, most recently, MSE R&D. As long as MSE remains vital for U.S. industrial and economic activity and as long as federal support of MSE R&D continues to be

[1]The breadth and depth of the contributions of MSE R&D to national priorities was demonstrated at two National Academies workshops: *Materials in the New Millennium—Responding to Society's Needs* (2001) and *Materials and Society—From Research to Manufacturing* (2003).

mittee that carried out the MSE benchmarking exercise added catalysts to the NSTC list and combined electronic materials and optical-photonic materials research into one category.[e] Since publication of the benchmarking report, the investigation of nanomaterials has blossomed into a vibrant, crosscutting area of materials research, so for the purposes of this report, the subfield of nanomaterials has been added to the list.

MSE at the beginning of the 21st century, therefore, consists of the following subfields:

- Biomaterials
- Ceramics
- Composites
- Magnetic materials
- Metals
- Electronic and optical-photonic materials
- Superconducting materials
- Polymers
- Catalysts
- Nanomaterials

In summary, MSE involves the generation and application of knowledge that relates the composition, structure, and processing of materials to their properties and uses.

[e]NRC, *Experiments in International Benchmarking of U.S. Research Fields*, Washington, D.C.: National Academy Press (2000).

a priority, the nation will need to understand well the drivers for and consequences of the globalization of MSE R&D.[2] The purpose of this report is to consider the issues surrounding this globalization. The report will

- Consider and identify trends in the evolution of MSE R&D being carried out around the world.
- Assess the forces that are driving globalization.
- Draw conclusions about the impact globalization might have on U.S. na-

[2]The effects and policy implications of the globalization of materials industrial activity, including the offshoring of U.S. manufacturing capacity are not considered in the report beyond looking at how such offshoring might affect decisions on locating R&D activities and the level of expertise in the United States.

tional priorities and, specifically, recommend actions to ensure the nation's continued access to current materials research.[3]

The approach of the Committee on Globalization of Materials Research and Development, which is responsible for this report, has been to identify trends in the globalization of MSE R&D based on evidence presented to it during the course of the study and on relevant reports and studies from other sources.

MATERIALS SCIENCE AS A PRIORITY FOR THE UNITED STATES

The critical task of setting policy for the nation's research endeavors has been the subject of many studies and reports. In 1993, the National Research Council (NRC) issued the report *Science, Technology, and the Federal Government: National Goals for a New Era*. In that report, the Committee on Science, Engineering, and Public Policy (COSEPUP) suggested that the United States adopt the principle of being among the world leaders in all major fields of science so that it could quickly apply and extend advances in science wherever they occur. In addition, the report recommended that the United States maintain clear leadership in fields that are tied to national objectives, that capture the imagination of society, or that have a multiplicative effect on other scientific advances. These recommendations were reiterated in another NRC report, *Allocating Federal Funds for Science and Technology* (1995), which said that the United States should "strive for clear leadership in the most promising areas of science and technology and those deemed most important to our national goals."

In 1999, the National Science and Technology Council (NSTC) stated that advanced materials were the foundation and fabric of manufactured products.[4] To support its assertion, NSTC cited the role of advanced materials in, among others, fuel-efficient automobiles, damage-resistant buildings and structures, electronic devices that transmit signals rapidly over long distances, protecting surfaces from

---

[3]While the expansion of MSE R&D activity and knowledge is closely connected to larger trends in the globalization of other economic activities—such as markets for goods and services and production networks—this report does not consider these except insofar as they drive or determine the location of MSE R&D.

[4]NSTC, *The Federal Program in Materials Science and Technology: An Overview Report*, 1999. The NSTC is a Cabinet-level council that is the principal means for the President to coordinate the diverse parts of the federal research and development enterprise.

wear and corrosion, and endowing jet engines and airframes with sufficient strength and heat tolerance to permit ever-faster supersonic flight. The NSTC concluded that many leading commercial products and military systems could not exist without advanced materials and that many of the new products critical to the nation's continued prosperity would come to be only through the development and commercialization of advanced materials.

In its report *Experiments in International Benchmarking of U.S. Research Fields* (2000),[5] COSEPUP asked, How important is it for the United States to lead in MSE? The materials subpanel that wrote the MSE-focused sections of that report noted that there had been an explosion in the understanding and application of MSE since the end of World War II, and that connections had become stronger between the materials field and other fields with emerging technology. The result, the subpanel concluded, was an acceleration in the contributions of materials to social advancement and economic growth.

A 1990 National Academies report, *Industrial Preparedness, National Resource and Deterrent to War*,[6] focused on how U.S. industry could be successful in an increasingly internationalized world. A key section, "Dependence on Overseas Sources," said

> The internationalization of the U.S. economy has caused global redistribution of the means of producing certain items. The United States is less self-sufficient than it was in past decades and is unlikely to be able to return to greater self-sufficiency. Because continual upgrades of technology are fundamental to a healthy domestic industry, federal incentives for technological investment will be more effective in the long run than "Buy American" restrictions.

The reports cited above represent only a small sample of many volumes that have been produced on the importance of materials research to future U.S. economic and national security and how the United States should react to the changing environment in which MSE R&D is taking place. They all point out that MSE research continues to address issues in agriculture, health, information and communication, infrastructure and construction, and transportation. Five areas are of particular interest.

---

[5]National Academy of Science, National Academy of Engineering, Institute of Medicine, *Experiments in International Benchmarking of U.S. Research Fields*, Committee on Science and Engineering Public Policy, Washington, D.C.: National Academy Press (2000).

[6]NRC, *Industrial Preparedness, National Resource and Deterrent to War*, Washington, D.C.: National Academy Press (1990).

TABLE 1.1 Historical Summary of DOD Materials Processing Science and Technology Funding (millions of current 2005 dollars)

Budget Item	1994	1995	1996	1997
Basic research	115.69	134.83	89.197	75.757
Applied research	366.77	386.99	316.38	279.18
Advanced Technology Development	28.434	25.368	37.33	29.026
Total budget request	510.9	547.2	442.9	384
Congressional add-on	147.39	120.5	35.728	48.176
Total appropriated budget	658.3	667.7	478.6	432.1

- The national defense of the country continues to depend on the ability to provide the most advanced weapons to the military, and the evolving threat to homeland security demands new materials to solve new problems.[7]
- MSE research continues to provide solutions to problems in health care with the development of new materials for the delivery of life-saving drugs and new implant technologies.
- MSE research is producing advanced materials solutions for more efficient energy production and transmission systems.
- MSE research is providing the latest materials for advanced transportation needs such as more energy-efficient and safer automobiles and advanced aerospace systems.
- Numerous consumer products benefit from MSE R&D.

Given the multifaceted importance of MSE R&D to the United States, maintaining world leadership in the field remains a critical national priority. This priority status can be evinced by the vibrant MSE R&D programs supported by federal agencies such as the National Science Foundation (NSF), the Department of Energy, the National Aeronautics and Space Administration (NASA), and the Department of Defense (DOD) and by the research programs of the various military services. While data are not available on aggregate federal support for MSE owing

[7]NRC, *Materials Research to Meet 21st Century Defense Needs*, Washington, D.C.: The National Academies Press (2003); NRC, *Accelerating Technology Transition: Bridging the Valley of Death for Materials and Processes in Defense Systems*, Washington, D.C.: The National Academies Press (2004); NRC, *Capturing the Full Power of Biomaterials for Military Medicine: Report of a Workshop*, Washington, D.C.: National Academies Press (2004); NRC, *Assessment of the Practicality of Pulsed Fast Neutron Analysis for Aviation Security*, Washington, D.C.: National Academy Press (2002).

1998	1999	2000	2001	2002	2003	2004	2005
101.28	85.978	86.219	111.99	96.06	113.14	102.84	83.8
274.3	337.42	283.8	306.93	322.67	292.88	299.93	264
32.968	37.71	40.05	47.589	47.6	43.806	33.871	36.3
384.8	461.1	410.1	466.5	466.3	449.8	436.6	384.1
38.423	70.43	79.544	100.2	80.265	114.29	153.5	168.5
423.2	531.5	489.6	566.7	546.6	564.1	590.1	552.6

to the highly diversified, interdisciplinary nature of MSE, some interesting trends can be seen in the recent budget data from NSF and DOD (Table 1.1 and Figures 1.1 and 1.2).

GLOBALIZATION AND R&D

What Is Globalization?

For such an extensively used term as globalization, it is notable that there is no precise and widely agreed definition. It has been suggested that globalization can be thought of as "the integration of free markets, nation states and information technologies to a degree never before witnessed and in a way that is enabling individuals, corporations and countries to reach around the world farther, faster, deeper and cheaper than ever."[8] The deepening integration that is characteristic of globalization is the source of both threats and opportunities. Globalization is a term that has recently taken on cultural, political, and other connotations in addition to the economic connotation. By and large, however, the most common or core sense of globalization remains economic and refers to the rapidly growing share of economic activity that now takes place between people, corporations, and institutions in different countries rather a single country. This growth in cross-border economic activity takes various forms—international trade, foreign direct investment, capital market flows, and so on. Each of these activities is associated with distinct issues, benefits, and risks that call for an assessment and a policy response.

[8]Thomas L. Friedman, "A Manifesto for the Fast World," *New York Times*, March 28, 1999.

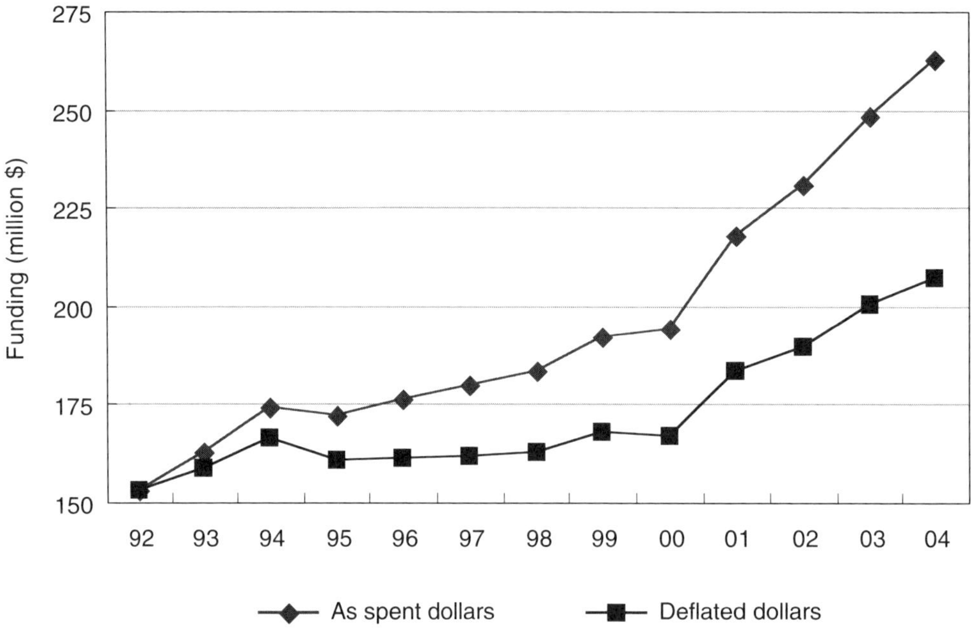

FIGURE 1.1 Funding trend at the NSF Materials Research Division (DMR). NSF/DMR experienced a budget increase in real terms of over 30 percent between 1992 and 2004. In FY2004 DMR's budget was close to $250 million. In the final FY2005 budget the only element to benefit from an increase was the $174 million for the Major Research Equipment and Facilities Construction account, up from $155 million in FY2004 but well short of the $213 million FY2005 request. These data indicate the generally upward trend in support for materials R&D at one division of NSF. Programs funded by other parts of NSF that might be thought of as being materials research—such as some chemical sciences programs and some engineering and nanotechnology programs—are not included here. Neither do these raw data show the details of R&D support by subdiscipline. Nonetheless the data from NSF/DMR give us insight, albeit limited, into current trends in federal funding of MSE R&D.

Notwithstanding the predominantly economic and trade-related features of globalization, in recent years a trend toward globalization of the world's R&D activities has emerged, driven by the institutions of government and by global industrial activity. The globalization of R&D can take a variety of forms—for example, offshoring industrial R&D activities; foreign direct investment in R&D in the United States and in other countries; and the development of transnational strategic alliances/cooperation, including intergovernmental supported networks of academic and industrial researchers.

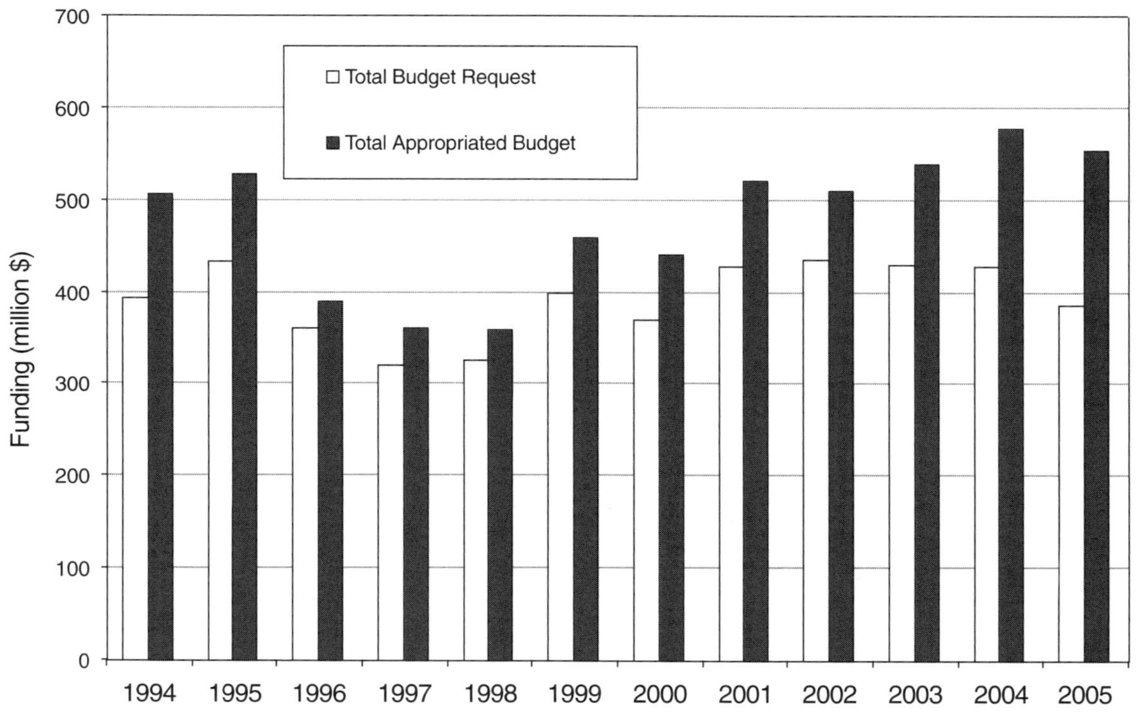

FIGURE 1.2 Historical summary of DOD materials processing science and technology funding (current 2005 dollars). Materials funding, also shown in Table 1.1, has been, on balance, flat in comparison with a decade ago. It is noteworthy that the final budget each year gained from congressional action to add funding—often for targeted projects—to the President's requested budget. SOURCE: Department of Defense.

What Is the Impact of Government?

The interplay between global corporations, national governments, and transnational organizations is multifaceted and complex. The globalization of industry and the increasing international competition between today's economies have caused some governments to plan and promote national capacities for the generation, absorption, and commercialization of new knowledge and the promotion of innovation. As R&D continues to become globalized, the policies of individual governments and regional groupings and organizations—such as the European Union (EU)—influence the flow of knowledge. This trend has led to new global centers for R&D emerging around the world in countries that might not have had a track record of large-scale R&D in the field in question.[9] Governments are recognizing the need for national and transnational responses to the issues globalization creates. Evidence of this can be found in the expanding importance of the EU's Framework Program, which now allows for the participation of researchers from outside the EU. In the United States the NSF is looking harder at the international perspectives.[10] In November 2004, NSF announced that the University of California, Santa Barbara, had been selected to host the International Center for Materials Research.[11] The mission of the center is to promote global

[9]An example of a new center of research can be found in Singapore's strategic development of biomedical research. Ernst and Young report in *Beyond Borders: The Global Biotechnology Report 2003* that the emergence of Singapore as a center for biomedical industrial activity—biomedical manufacturing output there increased by 48 percent in 2002, to US$ 5.5 billion—has been accompanied by a rise in R&D, with the emergence of global corporate R&D by companies such as Eli Lilly and Novartis, along with early-stage R&D by 25 Singapore-based corporations. As part of a case study on incentive programs in other countries, Ireland's Council for Science, Technology and Innovation (http://www.forfas.ie/publications/pharma_chem03/030526_icsti_pharmachem_statements.pdf) notes that expenditure on R&D in Singapore is deductible in the year in which it is incurred and that enhanced deductions are available for current expenditure on qualifying R&D (for example, double deductions for laboratory, testing, and medical research services). Capital allowances are available, including accelerated depreciation on R&D, computers, and other prescribed (for example, energy-efficient, pollution-reducing, and so on) equipment. The Technopreneurship Fund of over US$ 700 million provides industry-related training and offers participants international experience in key industries. The Singapore biomedical initiative was launched in 2000 with a US$ 750 million fund for investments in world-class biomedical firms undertaking R&D in Singapore. Specialist facilities being funded by the Singapore government include the Biopolis science park, which provides state-of-the-art laboratory facilities for biomedical sciences companies near research institutes and universities.

[10]See *Toward an International Materials Research Network*, available at http://www.nsf.gov/pubs/2002/nsf02068/background.htm.

[11]For more information, see http://www.icmr.ucsb.edu/general.html.

excellence in MSE through a series of research and educational programs and by providing an international forum for scientists and engineers with common interests in the future of MSE.

The emergence of a global innovation system appears to be intensifying the pressure for intergovernmental efforts aimed at the harmonization of regulatory-regime policy. As an example, at the 2004 ministerial meeting of the Committee for Scientific and Technological Policy[12] of the Organisation for Economic Co-operation and Development (OECD)[13] the ministers and senior government officials of 30 nations discussed and agreed on a declaration concerning access to research data from publicly funded research. Although the OECD declaration focuses on how to ensure open access to research data, it recognizes that the innovation system is becoming global, that this trend is beneficial, and that the flow of information is critical to supporting this trend. Governments are also restructuring national policies in order to attract R&D activities to their countries, often to underpin a wider effort to attract and maintain foreign direct investment in manufacturing and to build national R&D capacities. Box 1.2, "Building a Knowledge- and Innovation-Based Economy in the World's Most Globalized Country—Ireland," describes how Ireland is planning to support innovation in a globalized economy.

What Is the Impact of Industry?

The role of industry in MSE R&D appears critical to any discussion of the impact of globalization on MSE R&D. The irrefutably changing and global face of industry in the 21st century has resulted in the emergence of the global corporation. Indeed, globalization has arguably affected the management structures of many leading global corporations and how those corporations respond to pressure from stockholders to maximize profit. It is becoming commonplace that just as these corporations have manufacturing capacities around the globe, their ownership is often also spread among individuals, corporations, and institutions from

[12]The OECD committee aims at informing the policy debates on the contribution of science and technology to sustainable growth and societal needs in knowledge-based economies and at promoting international cooperation in scientific research. For information on its 2004 ministerial meeting, see http://www.oecd.org/cstp2004min.

[13]Current members of the OECD are Australia, Austria, Belgium, Canada, Czech Republic, Denmark, Finland, France, Germany, Greece, Hungary, Iceland, Ireland, Italy, Japan, Korea, Luxembourg, Mexico, Netherlands, New Zealand, Norway, Poland, Portugal, Slovak Republic, Spain, Sweden, Switzerland, Turkey, United Kingdom, and United States.

BOX 1.2 Building a Knowledge- and Innovation-Based Economy—Ireland

Many countries are formulating national innovation strategies not only in reaction to the increasing globalization of the R&D environment but also to exploit globalization with a view to supporting and developing opportunities for future economic growth based on high-technology industries. One such country is Ireland, which the A.T. Kearney/*Foreign Policy* Globalization Index ranks as the most globalized country in the world for the third year in a row.[a] Examining the Irish experience provides an insight into how one country is responding to globalization with a view to mitigating the risks and maximizing the potential.

Following two decades of concerted efforts by successive Irish governments to attract high-value investment by global high-technology industries, Ireland now enjoys living standards, as measured by gross domestic product (GDP) and gross national product (GNP) per capita, that exceed the EU average. Ireland's economy proved resilient through the global downturn in 2001–2003, although it grew more slowly than during the late 1990s. However, in recent years Irish policy makers have realized that future economic development would be strongly influenced by the shift toward services as an important driver of GDP growth and the increasing role of knowledge as a driver of economic development. In July 2003, Ireland's deputy prime minister, Mary Harney, summarized the challenge: "Put simply, in the past we have been excellent at making products designed by others; in the future we will have to design and innovate our own."

Ireland, in response, is undertaking a number of initiatives to create a new strategic orientation for Irish enterprise based on developing expertise in international markets; building a world-class research and innovation capability to support the development of high-quality, high-value products and services; renewing Ireland's commitment to education and training to provide the skills base industry will need; maintaining a competitive tax environment to drive economic growth; and providing capable and flexible government that can quickly identify and implement policies required to facilitate change.

Ireland has increased significantly its investment in science and technology. Under the implementation of Ireland's most recent national development plan for 2000–2006, the government is investing 2.54 billion in research, technological development, and innovation. The recently created Science Foundation Ireland (SFI) is investing 646 million be-

[a]*Measuring Globalization: Economic Reversals, Forward Momentum 2004*, A.T. Kearney/Foreign Policy Globalization Index. Available at http://www.foreignpolicy.com/story/cms.php?story_id=2493&page=0.

several countries. The 20 U.S. corporations spending the most on R&D (Table 1.2) are all highly globalized. The activities of corporations themselves within the global innovation system are important, and they can act in harmony with, against, or independent of national government policies.

Although no specific data for the international flow of MSE-related R&D funds are available, examining the broader transnational flow of R&D funds can tween 2000 and 2006 in academic researchers and research teams who are "most likely to generate new knowledge, leading edge technologies, and competitive enterprises in biotechnology, and information and communications technology."[b]

In executing its new R&D program, the Irish government has remained focused on developing a national innovation plan, an effort that has received the attention of the highest government leadership. At the first SFI science summit in September 2004, Ireland's Prime Minister, Bertie Ahearn, said, "the government's long-term vision is a thriving knowledge-driven economy built on truly competitive research. To achieve and sustain this goal, we must all work in partnership. That is, the researcher community working with their colleagues in education and industry for the greater good of our economy and our society."

A government report released in August 2004, *Building Ireland's Knowledge Economy*, lays out an ambitious vision:

> Ireland by 2010 will be internationally renowned for the excellence of its research and be at the forefront in generating and using new knowledge for economic and social progress, within an innovation-driven culture.

The plan laid out in the report goes beyond setting numerical goals for investment and acknowledges that what is required is no less than a transformation into "a pro-innovation culture" supportive of invention, risk-taking, and entrepreneurship. The strategy focuses on a systematic and continuous approach to encouraging R&D within firms. The government will also support the development of strategic research competencies (technology platforms) based on industry needs and develop seed capital markets for early-stage ventures. The strategy calls for a national plan to increase the performance, productivity, and efficiency of research in the higher education and public sectors, thereby sustaining Ireland's commitment to building an international reputation for research excellence. The program also aims to make Ireland a very attractive environment for high-quality researchers and research careers and develop the research commercialization expertise necessary to ensure effective and rapid exploitation by enterprise of research in the higher education and public research sectors.

[b]For further information on SFI programs, see http://www.sfi.ie.

provide some insight into the globalizing nature of the R&D environment. In the United States and in all the G-8 countries,[14] most of the funding for industrial R&D is provided by industry itself. Caution is needed when considering the data in

[14]The G-8 countries are Canada, France, Germany, Italy, Japan, the Russian Federation, the United Kingdom, and the United States.

TABLE 1.2 Top 20 R&D-Spending U.S. Corporations, 2001

Corporation	R&D Rank 2001	R&D Rank 2000	R&D Rank 1999	R&D (million $) 2001	R&D (million $) 2000	R&D (million $) 1999	Description
Ford Motor Company	1	1	1	7,400	6,800	7,100	Motor vehicle manufacturing
General Motors	2	2	2	6,200	6,600	6,800	Motor vehicle manufacturing
Pfizer Inc.	3	4	8	4,847	4,435	2,776	Pharmaceutical and medicine manufacturing
IBM	4	5	4	4,620	4,336	4,464	Computer systems design and related services
Microsoft	5	8	7	4,379	3,775	2,970	Software publishers
Motorola	6	3	5	4,318	4,437	3,438	Communications equipment manufacturing
Cisco Systems	7	11	20	3,922	2,704	1,594	Computer and peripheral equipment manufacturing
Intel	8	7	6	3,796	3,897	3,111	Semiconductor and other electronic component manufacturing
Johnson & Johnson	9	9	9	3,591	2,926	2,600	Pharmaceutical and medicine manufacturing
Lucent Technologies	10	6	3	3,520	4,018	4,510	Computer systems design and related services
Hewlett-Packard	11	12	10	2,635	2,646	2,440	Computer and peripheral equipment manufacturing
Merck	12	13	11	2,456	2,344	2,068	Pharmaceutical and medicine manufacturing
Bristol Myers Squibb	13	15	12	2,259	1,939	1,843	Pharmaceutical and medicine manufacturing
Eli Lilly	14	14	13	2,235	2,019	1,784	Pharmaceutical and medicine manufacturing
Pharmacia	15	10	25	2,195	2,753	1,290	Pharmaceutical and medicine manufacturing
Sun Microsystems	16	22	26	2,016	1,630	1,263	Computer and peripheral equipment manufacturing
General Electric	17	17	17	1,980	1,867	1,667	Engine, turbine, and power transmission equipment manufacturing
Boeing	18	24	22	1,936	1,441	1,341	Aerospace product and parts manufacturing
Wyeth	19	21	14	1,870	1,688	1,740	Pharmaceutical and medicine manufacturing
Procter & Gamble	20	16	15	1,769	1,899	1,726	Soap, cleaning compounds, and toilet preparations manufacturing

SOURCE: NSF, *Science and Engineering Indicators* (2004).

the paragraphs that follow, because a sizable portion of the R&D cited is likely to be for product development, including design, of products, or for short-range activities such as database generation.

The OECD reports that among its member countries, government financing accounts for a small and declining share of total industrial R&D spending—ranging from as little as 2 percent of industrial R&D performance in Japan to 11 percent in Italy.[15] In the United States in 2001, the federal government provided about 9 percent of the R&D funds used by industry, and the majority of that funding was obtained through DOD contracts. In recent years, the United States has attracted large investments by foreign companies that perform R&D in this country (Figure 1.3). This foreign-owned R&D grew at a real average annual rate of 10.8 percent from 1994 to 2000, mostly as a result of mergers and acquisitions, compared with a real average annual rate of 6.9 percent for U.S.-owned R&D overseas.[16] Foreign-owned firms conducting R&D in the United States accounted for $26.1 billion (13 percent) of the $199.5 billion in total industrial R&D expenditures in the United States in 2000.[17] This share fluctuated between 11 and 13 percent from 1994 to 2000. Seven countries invested $1 billion or more in R&D in the United States in 2000: Germany, the United Kingdom, Switzerland, Japan, Canada, France, and the Netherlands, accounting for about 90 percent of all R&D expenditures in the United States by foreign-owned firms.[18]

Two-thirds of all R&D performed overseas in 2000 by U.S.-owned companies ($13.2 billion of $19.8 billion) took place in six countries: the United Kingdom, Germany, Canada, Japan, France, and Sweden. At the same time, emerging markets such as Singapore, Israel, Ireland, and China were increasingly attracting R&D activities by subsidiaries of U.S. corporations. In 2000, U.S.-owned R&D expenditures reached $500 million or more in each of these emerging markets, considerably more than in 1994. OECD reports that in the 20 years from 1981 to 2001, the increasing globalization of industrial R&D activities resulted in a concomitant increase in the share of foreign sources of R&D funding in many countries (Figure 1.4). The role of foreign funding in R&D varies from country to country, account-

---

[15]OECD, *Main Science and Technology Indicators* (2002).

[16]NSF, *Survey of Industrial Research and Development* (2003). Available at http://www.nsf.gov/sbe/srs/indus/start.htm.

[17]Source: NSF *Indicators* report. Based on the estimated percentage of U.S. industrial performance undertaken by majority-owned (i.e., 50 percent or more) nonbank U.S. affiliates of foreign companies. Foreign R&D investments in the United States represent industry funding based on foreign ownership regardless of originating source, whereas the foreign totals for other countries represent flows of foreign funds from outside the country to any of its domestic performers.

[18]In 2000, European-owned subsidiaries accounted for $18.6 billion (71 percent) of foreign-owned R&D in the United States, a share comparable with their 67 percent share in foreign-owned gross product in the United States. The corresponding R&D shares for Canadian- and Asia/Pacific-owned subsidiaries were 14.0 and 10.9 percent, respectively. In particular, R&D activities by U.S. affiliates of foreign companies were dominated by seven investing countries with $1 billion or more in R&D expenditures.

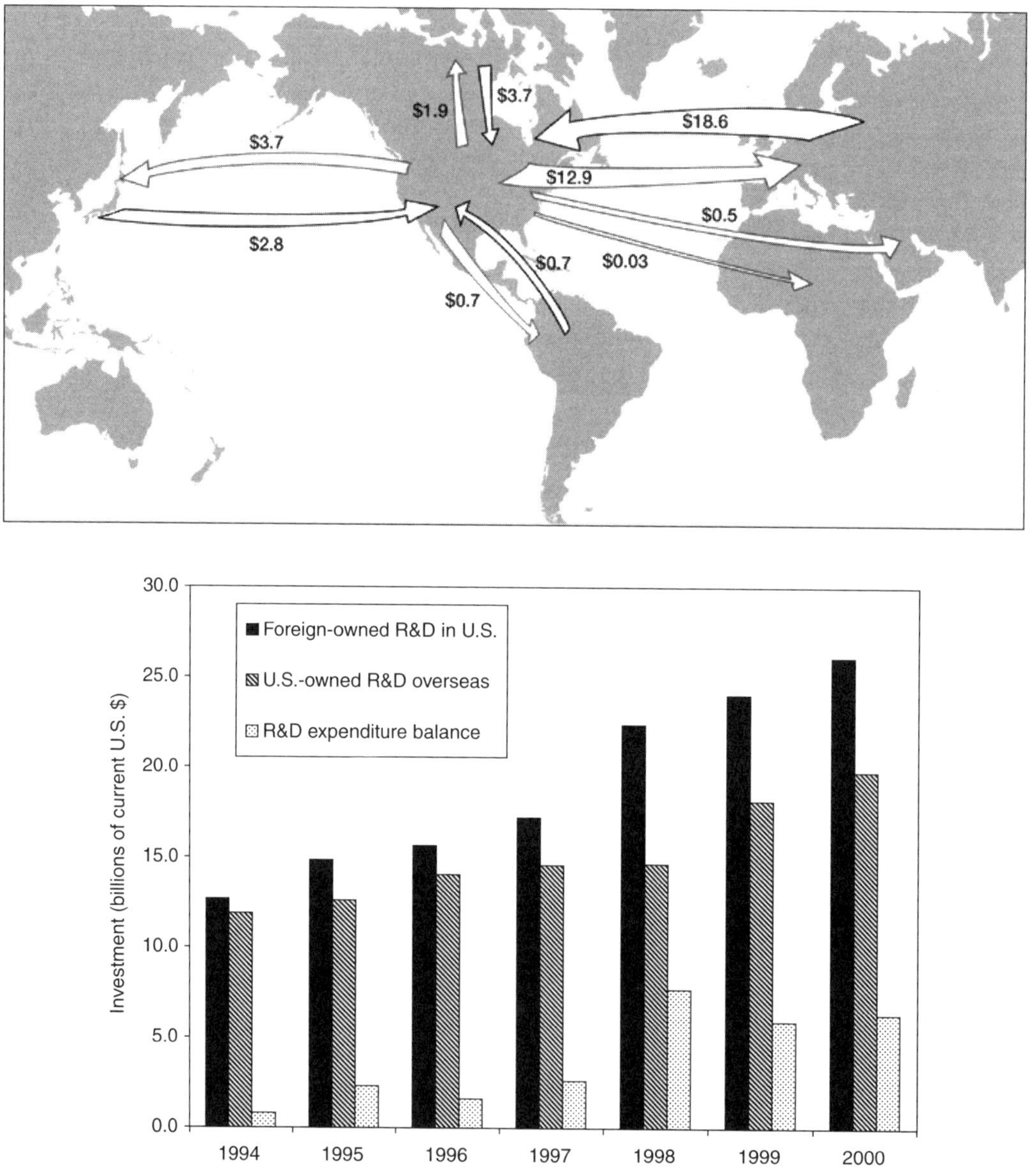

FIGURE 1.3 Foreign-owned R&D in the United States and U.S.-owned R&D overseas. Upper chart shows the flow of R&D investment into and out of the United States and the lower chart shows the trend of investments between 1994 and 2000. The European region accounted for approximately two-thirds ($12.9 billion) of all U.S-owned overseas R&D. The Asia/Pacific region ($3.7 billion, or 18.9 percent) outpaced Canada ($1.9 billion, or 9.5 percent) as a locale for U.S.-owned overseas R&D. SOURCE: NSF, *Science and Engineering Indicators* (2004), available at http://www.nsf.gov/sbe/srs/seind04/start.htm.

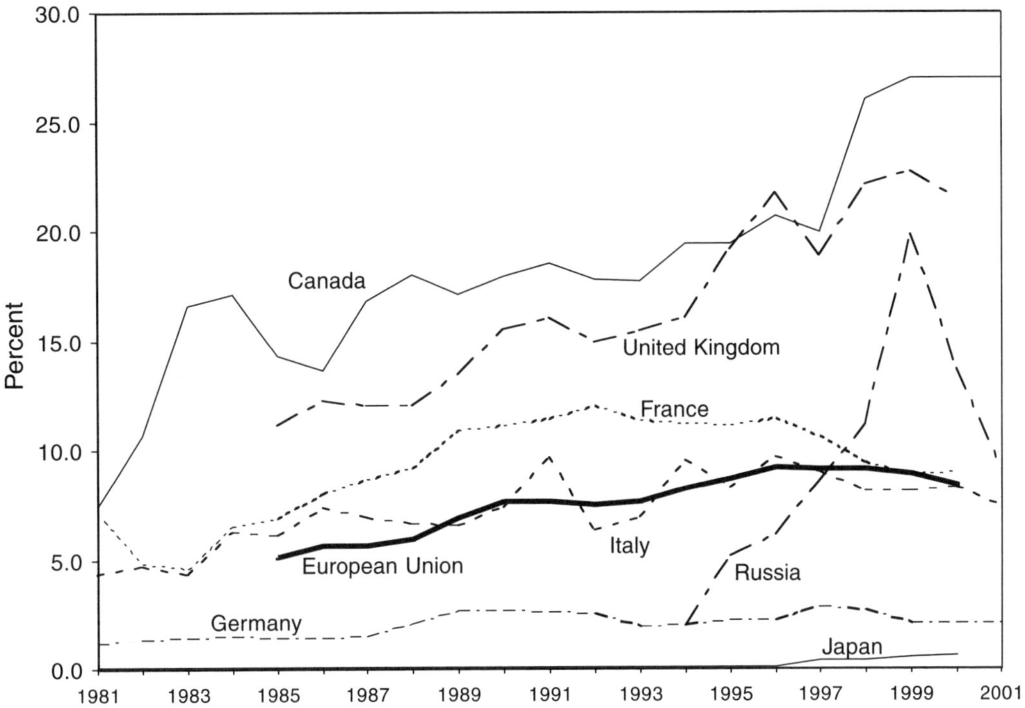

FIGURE 1.4 R&D financed by foreign sources as percentage of total industrial R&D, 1981–2001. Foreign sources of R&D funding increased in many countries between 1981 and 2001. This foreign funding came predominantly from foreign corporations but also from foreign governments and other foreign organizations. SOURCE: OECD, *Main Science and Technology Indicators* (2002).

ing for as little as 0.4 percent of industrial R&D in Japan to as much as 27 percent in Canada in recent years. Foreign funding as defined by the OECD predominantly came from foreign corporations but also included funding from foreign governments and other foreign organizations.[19] According to a recent survey of senior executives, over half the leading companies polled plan to increase their overseas R&D investments over the next 3 years and a further 38 percent will

[19]The expansion of efforts through successive EU Framework Programs to foster cooperative research throughout the EU may have added to the growth in foreign sources of R&D funds within Europe.

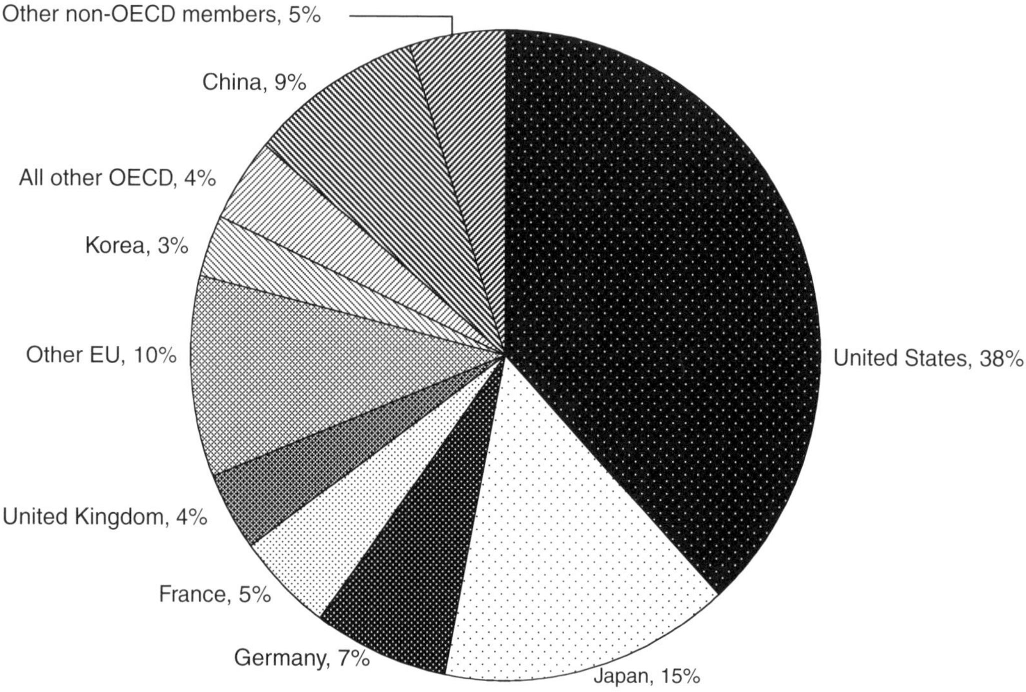

FIGURE 1.5 The United States continues to lead the world in R&D, with 38 percent of world R&D spending in 2002, according to data from the OECD. U.S. industry, government, and other sectors spend more on R&D than the entire EU combined. The U.S. share has declined only slightly from 40 percent during most of the 1990s, but other nations' shares have changed considerably. China has increased its R&D investments dramatically in recent years and is now the third largest investor in R&D (adjusted for purchasing power), behind only Japan and the United States. South Korea is the only other country that accounted for a substantial share of the OECD total (3.1 percent in 2000, which was higher than the share of either Canada or Italy). In only four other countries (the Netherlands, Australia, Sweden, and Spain) did R&D expenditures exceed 1 percent of the OECD R&D total. SOURCE: OECD, *Main Science and Technology Indicators* (2004). Data gathered for 2003 or latest year available and calculations based on purchasing power parities.

sustain existing spending levels.[20] Although data exist on nondomestic investments of R&D funding flowing into other countries, there are no equivalent data on foreign investments in R&D in this country. However, the importance of foreign investment in U.S. R&D is shown by the fact that approximately 13 percent of

[20]Economist Intelligence Unit, *Scattering the Seeds of Invention: The Globalization of Research and Development*, September 2004.

TABLE 1.3 International R&D Expenditures and R&D as a Percentage of GDP, for Selected Countries and for All OECD Countries, 1981–2001

	United States	Japan[a]	Germany[b]	France	United Kingdom	Italy	Canada	Russian Federation	Total OECD
R&D expenditure (billions of constant 1995 U.S. dollars)[c]									
1981	114.5	39.7	27.8	17.4	18.2	7.7	6.0	NA	254.9
1986	157.3	55.6	33.5	21.9	20.1	10.9	8.2	NA	339.6
1991	176.6	77.6	42.0	27.2	20.6	13.4	9.7	20.5	415.7
2001	252.9	NA	48.6	30.7	NA	NA	16.2	10.9	NA
R&D as share of GDP (%)									
1981	2.34	2.11	2.43	1.93	2.38	0.88	1.24	NA	1.95
1986	2.73	2.51	2.70	2.21	2.26	1.13	1.48	NA	2.26
1996	2.55	2.77	2.26	2.30	1.88	1.01	1.69	0.90	2.13
2001	2.82	NA	2.53	2.20	NA	NA	1.94	1.16	NA

[a]Data on Japanese R&D in 1996 and later years may not be consistent with data in earlier years because of changes in methodology.
[b]Data for 1981–1990 are for West Germany.
[c]Conversions of foreign currencies to U.S. dollars are calculated with each country's GDP implicit price deflator and with OECD purchasing power parity exchange rates.
SOURCE: OECD, *Main Science and Technology Indicators* (2002).

funds spent on industrial R&D in 2000 is estimated to have come from majority-owned affiliates of foreign firms investing domestically.[21]

Notwithstanding the globalizing trend, it is interesting to note that, worldwide, R&D performance remains concentrated in only a few industrialized nations. The OECD reports that of the $603 billion in estimated R&D expenditures for the 30 OECD countries in 2000, 85 percent is expended in only 7 countries (Figure 1.5). While in absolute numbers the United States is dominant among R&D spenders, analyzing expenditure as a fraction of economic activity provides an interesting perspective (Table 1.3 and Figure 1.6). OECD reports that between 1995 and 2002 China doubled its R&D spending, from 0.6 percent to 1.2 percent of GDP.[22] By comparison, overall R&D spending in OECD countries rose only

[21]The figures used here to approximate foreign involvement are derived from the estimated percentage of U.S. industrial performance undertaken by majority-owned (i.e., 50 percent or more) nonbank U.S. affiliates of foreign companies. The U.S. foreign R&D totals represent industry funding based on foreign ownership regardless of originating source, whereas the foreign totals for other countries represent flows of foreign funds from outside the country to any of its domestic performers.

[22]OECD, *Science, Technology and Industry Outlook* (2004). Available at http://www.oecd.org.

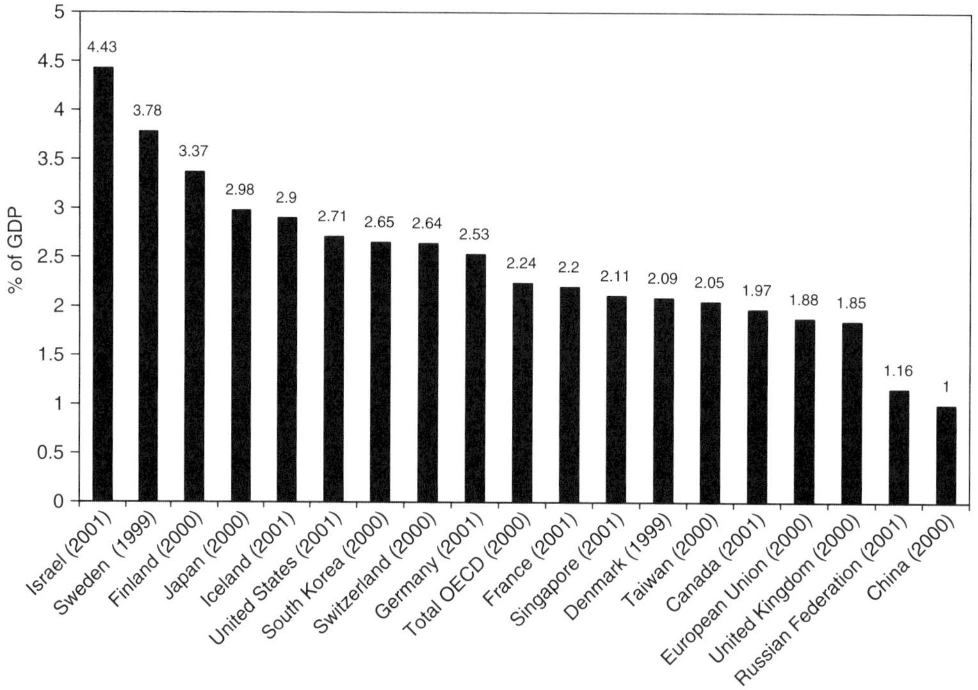

FIGURE 1.6 R&D expenditure as a percentage of GDP. Overall, the United States ranked fifth among OECD countries in terms of reported R&D/GDP ratios and sixth when Israel (not an OECD member country) is included. Israel, devoting 4.4 percent of its GDP to R&D, led all countries, followed by Sweden (3.78 percent), Finland (3.37 percent), Japan (2.98 percent), and Iceland (2.9 percent). SOURCE: OECD, *Main Science and Technology Indicators* (2002).

moderately in proportional terms, to 2.26 percent of GDP from 2.09 percent in 1995. In 2002 the R&D expenditure of all 25 EU member states stood at 1.93 percent of GDP, an increase from 1.82 percent in 1998. In the United States, R&D performance as a proportion of GDP rose, from 2.40 percent in 1994 to 2.69 percent in 2000, as growth in R&D outpaced the growth of the overall economy. Looking ahead, a number of countries have set long-term targets for increasing R&D spending, with Austria aiming for 2.5 percent of GDP by 2006, Germany 3.0 percent by 2010, and the United Kingdom 2.5 percent by 2014. Canada has set for itself the target of being among the top five OECD countries with respect to investments in R&D, and Korea has committed to doubling its investment between 2003 and 2007.

OECD countries accounted for 97.6 percent of patent applications to the European Patent Office (EPO) in 1999 and for more than 95 percent of patents granted by the U.S. Patent and Trademark Office (USPTO) in 1998. In 1999, Israel—at 122 patent applications per million population—was the only non-OECD economy whose patent applications at the EPO exceeded the OECD average of 88. In 1998, Taiwan had 223 patents granted at the USPTO per million population. Non-OECD economies accounted for only 1.5 percent, up from 1 percent in 1991.

2

Indicators for Trends in Globalization

IS MSE R&D BECOMING A GLOBAL ACTIVITY?

Chapter 1 considered globalization and its impact on R&D as well as discussing the scope and definition of MSE. Researchers, engineers, and scientists active in MSE R&D believe that their field is becoming increasingly globalized. But what data are available to back up this common wisdom? Trends in global R&D activity can be difficult to demonstrate with clarity. Typically, when analysts try to identify and assess the internationalization or globalization of R&D, they turn to patent data, trends in the national origins of literature in major scientific journals, education trends, the activities of professional societies, and trends in the activities of international corporations. For this report the committee uses the same methodologies. First, however, it defines the globalization of MSE R&D as the worldwide expansion of MSE knowledge-creation centers as a result of U.S. and non-U.S. industry and government investments along with the worldwide increase in collaboration facilitated by information technology.

The Use of Patent Indicators

As a policy instrument, patents are intended to encourage innovation by providing a framework for protecting intellectual property. Patent data can be mined for indicators that measure the output of R&D in particular sectors. Patent data have advantages and disadvantages. Among the advantages are the wealth of information contained in patent records—on the applicant, the inventor, locations, and so on. However, the interpretation of patent data has many caveats associated with it. For instance, the value distribution of patents is skewed because many patents awarded have little or no industrial application.[1] In addition, many inventions are never patented—corporate inventors, for instance, might choose to keep them as trade secrets for competitive advantage. Also, differences in national intellectual property (IP) regimes can hinder the accurate disaggregation and interpretation of data. Changes in patent law and in patenting patterns make analysis over time difficult. The time lag between an invention and the award of a patent for that invention also causes patent data analysis to be insensitive to very recent trends in invention. Nevertheless this tracking of international activity in a particular field is a good complement to other methodologies and can identify trends.

Because the focus of this report is to look at the globalization of MSE R&D, information on the inventor's country of residence at the time of the invention is of paramount interest and can indicate shifts in the international location of inventive activity. Although patents with inventors from more than one country are likely to have been counted more than once, the trends displayed are unlikely to change if fractions of those patents are allocated across the countries.

Before looking at the indicators for some key MSE subfields, it is worth asking if there are overall trends in patenting that can help identify baseline trends for patenting in general and from which we can discern MSE-specific trends. These overall trends in all patents are reported in more detail in Appendix C and can be summarized as follows. Fifty-three percent of the inventors associated with U.S. patents reside in the United States, more than in any other country and a larger share than that of the European Union and Japan combined. However, caution is warranted and advisable when drawing any conclusions about the U.S. share of R&D activity based on patent share because of what is called the "home advantage," whereby proportionate to their inventive activity, patent applicants are more likely to file for a patent in their home country than to file with a foreign patent office. The Organisation for Economic Co-operation and Development (OECD) has worked hard to understand the home advantage, and the data it has amassed point to an increase in transnational activity—that is, a growing share of patents in any given country is owned by individuals, companies, or organizations residing in other countries, indicating an increase in global and transnational ownership and collaboration.

---

[1]M. Schankerman, "How Valuable Is Patent Protection? Estimates by Technology Field," *RAND Journal of Economics* 29 (1998): 77–107; A. Pakes, "Patents as Options: Some Estimates of the Value of Holding European Patent Stocks," *Econometrica* 54 (1986): 755–784.

Global Trends in MSE Patent Data

While the OECD has analyzed data on patents to identify trends for aggregate R&D, no similar analysis has been carried out for patents relating to MSE. The committee therefore carried out a limited and focused analysis of the location (by country) of inventors associated with U.S. patents awarded over the last 25 years in certain key and well-defined MSE subfields. Notwithstanding the weaknesses associated with single-patent-office (USPTO) methodology, such an analysis can be useful in understanding the extent to which U.S. dominance has changed over time, a key question for this study. The data were gathered from the online USPTO database.[2]

Patents of interest were identified by their associated World Intellectual Property Organization (WIPO) category or by searching for a keyword in the patent title. The number of inventors from countries with well-established and emerging levels of activity in MSE were recorded. The key subfields chosen for this analysis were alloys, catalysts, ceramics, magnetic materials, and composites.

It is worth noting that since the issue at hand is the level of global activity in MSE R&D, the information tracked by the committee was the location of the inventors recorded on the patent approval. The following analysis does not record whether the work that led to the invention was part of an international collaboration or partnership or some other form of transnational arrangement. In addition, because the main purpose in carrying out these analyses is to understand the evolution of U.S. and non-U.S. R&D activity over the period, the data were normalized to the numbers for the United States. The following summarizes the results in each of the subfields. A fuller presentation of the results can be found in Appendix C.

Alloys

Patent applications in the alloys subfield are dominated by inventors in the United States, Japan, and Western Europe. U.S. activity remained fairly steady from 1979 to 2004, at around 550 patents a year. Japan significantly increased its absolute number of patents (from 251 to 653 in the period reported), and its share (relative to that of the United States) surged, surpassing the U.S. share in the mid-1990s. Western Europe has had a steady increase in activity, with its share relative to the U.S. share increasing by 50 percent over the last 25 years. While the number of patents awarded with inventors in China and Korea—the two strongest nontraditional performers—has increased, the absolute numbers remain an order of

[2]The search engine on the European Patent Office (EPO) Web page did not allow for a similar data mining exercise.

magnitude behind Europe, Japan, and the United States. Activity rose sharply, however, and may reflect early and emerging stages of globalization of this type of R&D in Asia.

Catalysts

Patents for catalysts exhibit some different trends. The dominance of the United States is persistent, with over twice as much patent activity for inventors in the United States as for inventors in Europe and Asia. However, the numbers for inventors in Germany and Italy have surged, and France remains a strong participant. Among the emerging centers of R&D, China, India, Korea, and Taiwan all show substantial increases over 25 years ago. As is the case with the alloy data, the increase in these countries is strong, but the number of patents remains an order of magnitude less than in Japan and Western Europe and two orders of magnitude less than in the United States.

Ceramics

Patents in ceramics are dominated by the United States and Japan. The number of patents with inventors in Japan jumped significantly at the beginning of the 1980s, and activity there recently appeared to be on a par with the United States. Given the home advantage for the United States, Japan may have equaled or even surpassed the United States in the last decade. Germany dominates European activity, but its share remained fairly steady over the last 25 years. Taiwan and Korea show the strongest increases in performance, but the absolute numbers remain very low.

Magnetic Materials

In magnetic materials, patent output from the United States and Japan has grown along with economic activity and outperformed the combined activity of the five European countries by a factor of 4. Europe's share decreased over the past 20 years, primarily because its output stagnated when that of Japan and the United States was growing. The data also show significant increases in output from Taiwan and Korea—although as with the other subfields, the absolute numbers of patents with inventors in those countries remains low.

Composite Materials

In the field of composite materials there has been a noticeable increase in global research, with patent output from the United States, Asia, and Europe about equal. Activity in Europe is dominated by Germany and France. Patent output by inventors in Italy shows a significant upward trend, while activity in the United Kingdom and Switzerland remains static. The United States appears to have lagged behind Japan in the mid-1980s but has caught up since. Taiwan and Korea have been active, but overall numbers remain low; China and India have displayed no significant activity.

Summary of Patent Data

Although there are significant differences in the indicators for the five subfields measured and although there are limitations to this type of analysis, some general trends emerge:

- In each subfield the United States is the world leader or among the world leaders.[3]
- Japan and Western Europe are the closest rivals for leadership. Japan appears to have surpassed the United States in alloys and seems set to surpass it in ceramics.
- Global activity in all the subfields examined is diversifying, with significant increases in activity in Asian countries that were not previously very active in these fields. How this trend may evolve is unclear.
- Notwithstanding the global diversification of research activity, the number of patents with inventors in the emerging centers remains low relative to patents by U.S. inventors.

Global Trends in MSE Literature Data

While trends in patenting help to elucidate global innovation, their ability to track basic research might be called into question. Trends in the authorship of journal articles around the world should be expected to reflect broader R&D activity in science and engineering in a manner complementary to trends in patenting. Accordingly, the location of authors has been searched for five key subfields of materials research. As research grows around the world, the numbers of authors from countries with an emerging presence in materials should grow. Details of this information can be found in Appendix D. However, it is important to point out that this kind of analysis does not measure the quality of the research or its impact on the relevant field.

---

[3]Subject to the qualification that it is patents filed in the United States that are being counted.

Alloys

In the alloys subfield, the global output of scientific papers grew considerably in the past 20 years. The number of papers from the United States increased steadily, but its share of the global output has decreased. The same is true for Japan, whereas Europe and Asia steadily increased their share of the global output. In particular, the number of papers by authors from China and Korea shows exceptional growth, with the number from China approaching the numbers from the United States and Japan and outstripping traditional R&D leaders such as Germany, France, and Britain.[4]

Catalysts

In catalysis there was an increase in the number of papers over the past 20 years. Notwithstanding the strong increase in papers authored in the United States, the U.S. share of the global total declined somewhat after 1990. The United States does not enjoy any definitive lead but remains among the leaders. Japan's share has remained steady, and the numbers from the Euro5 show a small but discernible increase in the 1990s but some stagnation more recently. The Asian share of the literature output showed a steady and strong increase over the last 20 years, outstripping U.S. activity over the last 5 years. China shows a particularly strong increase, surpassing the individual European countries and approaching the output of Japan. Korea and, to a lesser extent, Taiwan show significant increases in share.

Composite Materials

In composite materials there has been a surge in global output over the last 15 years, with the United States remaining in the lead. However, the data also show this leadership being challenged recently by the European and Asian regions. In Europe, Britain enjoys the lead, with Germany close behind. China's share is approaching Japan's, and significant increases can be seen in the share of output for Taiwan and Korea, although the numbers are small.

---

[4]Literature data were collected for Great Britain (England, Scotland, and Wales) while patent data were collected for the whole of the United Kingdom (including Northern Ireland). This inconsistency was due to the configuration of the search engines of the respective databases.

Optical-Photonic Materials

In optical-photonic materials, the United States maintained its lead, but over the last 15 years the European and Asian regions have been challenging the United States. Italy shows the strongest surge in Europe. Japan's share remained steady over the 20 years, although the increases seen in China, Taiwan, and Korea were not as strong as in the other subfields.

Summary of Literature Data

Although there are differences from one subfield to another, some trends emerge:

- The United States is among the world leaders in all four subfields, but it does not clearly dominate in any of them.
- Western Europe and Japan are also among the leaders in these subfields, showing clear surges in activity over the last 15 years, overcoming previous U.S. dominance.
- In all four subfields there have been significant increases in the literature presence of the Asian countries included in these searches; most notably, the number of papers from China increased substantially over the last 5 to 10 years, at times approaching the number from Japan and several West European countries.
- Authorship in the world's journals appears to be globalizing rapidly, with clear gains for authors from countries that had no traditional background in these fields.

GLOBALIZATION OF THE U.S. MATERIALS COMMUNITY

To understand better the globalization experienced by the materials community, a Web-based poll was conducted over a 2-week period. An e-mail announced the poll to members of the materials professional societies. The responses to questions in the poll are summarized in Appendix E. It is important to bear in mind that the exercise was based on a self-selected response group. In addition, while a broad range of societies (see Appendix E) assisted in this exercise, reflecting the broad nature of MSE, it cannot be assumed that the self-selected group fully represents the increasingly interdisciplinary MSE community. (See Box 2.1, "Recent Experience of MSE Professional Societies.") The poll should therefore be considered as providing qualitative information on the global nature of MSE activity. A more reliable poll would require applying the usual criteria in choosing a statistically relevant sampling.

BOX 2.1 Recent Experience of MSE Professional Societies

The general health of the field may be gauged by the number of professionals staying active in MSE. Table 2.1.1 shows membership trends for the professional societies that most materials scientists and engineers identify with.[a] The trend shows clearly a loss of about 25 percent in membership between 1996 and 2003, but there are details that are of more concern. For example, while the American Ceramic Society (ACerS)—the home of the National Institute of Ceramic Engineers (NICE)—has about 8,000 members, fewer than 1,000 are members of NICE, and many ACS members are ceramic artists. Also, while ASM International (the former American Society for Metals) has about 34,000 members, only about 27,000 are active professionals and around 7,000 are retired senior members. The average age of members of ASMI is 48. The median age of the 6,500 members of The Minerals, Metals and Materials Society (TMS) is 53, and 1,200 are senior members over 65. There are about 1,000 student members jointly in these two societies. When joint memberships of both societies are taken into account, the total active is only about 29,000. The shares of foreign members in ASMI and TMS are 18 and 33 percent, respectively. The membership of the Society of Manufacturing Engineers, which obviously represents a much broader set of interests, also fell 30 percent over the 1996–2003 period.

TABLE 2.1.1 Materials-Related Professional Engineering Society Memberships

Society[a]	1996	1997	1998	1999	2000	2001	2002	2003 (est.)
ASM	43,317	43,329	42,264	40,407	38,227	36,656	34,450	34,000
ACerS	10,324	10,340	10,306	9,803	9,347	8,646		
TMS	11,236	11,532	8,596	8,646	7,375	7,067	6,487	6,500
SME	66,800	63,900	61,500	55,741	52,082	49,078	46,703	45,000
SPE	38,287	37,748	37,164	29,524	28,618	24,985	21,237	19,475
ISS	8,061	8,223	8,551	7,844	9,373	7,473	6,071	6,000

[a]ASM, American Society for Metals, now ASM International; ACerS, American Ceramic Society; TMS, The Minerals, Metals and Materials Society; SME, Society of Manufacturing Engineers; SPE, Society of Plastics Engineers; and ISS, Iron and Steel Society.

[a]The data on society membership were gathered in private interviews with representatives of the organizations.

Notwithstanding this caveat, the makeup of the response group was varied. A total of 719 respondents completed the questionnaire between November 8 and 13, 2004, with 144 identifying themselves as being based outside the United States. The respondents were drawn from a wide variety of organizations—U.S. universities, 32.9 percent; U.S.-headquartered corporations, 27.7 percent; U.S. national laboratories, 14.6 percent; non-U.S. universities, 9.7 percent; overseas-

TABLE 2.1 Nature of International Collaboration

Type of Collaboration	Share of All Collaborations (%)
U.S. academic-foreign academic research	54.5
U.S. corporate-foreign corporate research	14.1
U.S. academic-foreign corporate research	6.5
U.S. corporate research carried out by foreign affiliates of the U.S. corporation	12.3
U.S. corporate research carried out with joint ventures and/or by contract with foreign corporation(s)	12.6

NOTE: Results of questionnaire sent to MSE researchers who self-identify as being in the United States and carrying out research with an international aspect. (These data are indicative only and not based on a statistically relevant sampling.)

headquartered corporations, 6.7 percent; non-U.S. national laboratories, 2.1 percent; and other, 6.3 percent. Thirty-five percent of the respondents reported that they carried out their research exclusively in the United States, 54.8 percent carried out their research mostly in the United States, and 10.3 percent reported carrying out their research in an international partnership.

Respondents who reported some international element to their research activities were asked to clarify the international nature of their work (Table 2.1).

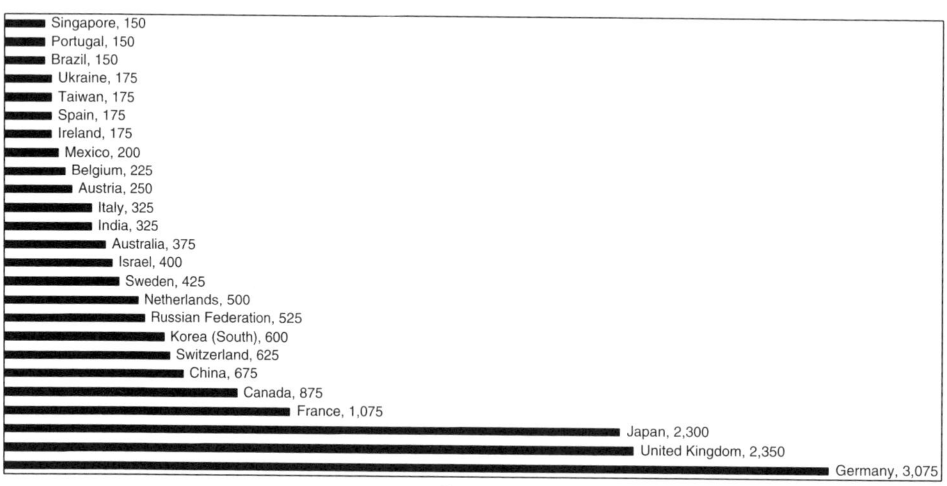

FIGURE 2.1 Respondents to a poll of the MSE community in this country were asked to list their top three international partners. The committee weighted the first choice by an arbitrary score factor of 100, the second choice by 50, and the third choice by 25. The rankings above are based on the total weighted scores. International partnerships were dominated by activity with Germany, Japan, and the United Kingdom. The rankings are indicative only and not based on a statistically relevant sampling.

Germany was identified as the most important partner by a majority of the respondents in each ranking. Japan and the United Kingdom also remain important partners according these data, and China, Korea, and Russia are just as important or perhaps more important partners than more-traditional European partners (see Figure 2.1).

It is important to recall the potential bias of the self-selected response group. Nevertheless the responses indicate that about one-third of U.S. MSE research appears to be carried out exclusively in the United States, and the international activity reported is dominated by research carried out mostly in the United States. The majority of international activity reported is between researchers at academic institutions. Twenty-four percent of U.S. research reported is carried out between U.S. corporations and foreign affiliates of U.S. corporations or in joint arrangements with foreign corporations.

In summary, MSE R&D in the United States is an internationalized activity with a diversified set of international partners.

INFORMATION TECHNOLOGY AS AN ENABLER OF GLOBALIZATION

Technology available in the marketplace today has enabled the globalization of all R&D areas, including MSE R&D. Advances in information technology and communications technology have made current research information accessible to even the most remote parts of the world. Cell phones are inexpensive, and cell phone service is available in most populated areas. Communication via satellites is more costly yet not prohibitively so and offers broadband service. Communication, whether wireless or land based, is an essential component of getting connected to the World Wide Web. According to the A.T. Kearney/*Foreign Policy* Globalization Index, more than 130 million new Internet users came online in 2002, bringing the total to more than 620 million. Arguably, the Internet has had the greatest impact on the globalization of research.

The Web has affected what research people perform and how and where they perform it. Internet search engines can supply researchers at their desks with information on active research areas, research needs, published information, how the research is performed, and potential places for collaboration. Of course the expansion of the Internet has coincided with the proliferation of inexpensive, high-performance computers. Computer processors continue to increase in power yet decrease in price. The Internet is the vehicle for e-mail, teleconferencing, videoconferencing, and telephony. E-mail is a powerful communication method and can support attachments such as documents, photos, and audio/video clips. People around the globe have come to accept the underlying technologies that allow Internet connectivity—communication protocols, digital formats, interconnects, viewers, and common application software. The Web and e-mail offer a faster, cheaper, and better method of obtaining and transferring information.

The Internet and e-commerce have facilitated the global availability of research tools—for example, instrumentation, laboratory equipment, computational computers, and modeling software—allowing research methods to be mimicked anywhere in the world. The current generation of computer-controlled laboratory testing equipment greatly enables the diffusion of materials R&D to developing countries. Today's scanning electron microscopes, gas-chromatography-coupled mass spectrometers, Fourier transform infrared microscopes, and ion chromatographs use Windows-based computer control and can be networked to the Internet. Not only do you not need a Ph.D. to operate these instruments, but one trained technologist can operate many different machines and share the results globally over the Internet for evaluation and discussion. The availability of these machines encourages locating advanced materials labs at major manufacturing sites around the world to provide rapid onsite evaluation and development.

In summary, information technology and global communications are affecting the execution of R&D in innumerable ways and enabling R&D to evolve new and global modalities.

GLOBALIZATION OF CORPORATE MSE R&D

Offshoring is part of the global trade system, and the decision to offshore corporate activity is a strategic business decision. Offshoring no longer involves just manufacturing but also extends to other functions, including services, from human resources and information technology management to R&D. Over time, shifts in manufacturing capabilities can result in the migration of research, design, and development activities—see Box 2.2, "Globalization and an American Company: The Timken Company." Technology transfer is also becoming a global phenomenon—see Box 2.3, "An Evolving Model of Technology Transfer: The Transfer of Nanotechnology from New Zealand to an American Firm for Introduction to the Global Market." Recent actions announced by a number of corporations reaffirm the globalized marketplace and continuing offshoring trends—see Box 2.4, "Recent Corporate Global Announcements."

What are the drivers of decisions to globalize corporate MSE R&D portfolios? While it appears that no data specific to MSE are available, it is reasonable to assume that the trends in transnational corporate MSE R&D are not significantly different from trends for all R&D, recognizing that specific materials subfields may exhibit more intense trends or trends that are different from those of the field as a whole. It is also reasonable, therefore, to assume that the major concerns of corporations involved in MSE R&D are similar to those of corporations involved in

R&D in other sectors: that is, the continuing globalization of the knowledge economy, including competition from other countries to attract innovation activities; the importance of being close to customers, be they end users or manufacturers; global trends in education, manufacturing, and the workforce; increasing global competitiveness; attractiveness of alternative regulatory and tax environments; and the need to access and tailor products to local markets. One of the most recent surveys in this regard was carried out by the Economist Intelligence Unit (EIU).[5]

In the EIU survey of 104 corporate leaders from a broad range of leading global companies, 71 percent of executives cited the ability to exploit pools of skilled labor as a key benefit of globalizing R&D. Other important benefits identified in the surveys included the ability to tailor goods and services to particular markets, reduced R&D costs, and reduced time to market. When asked to rate which aspects of the business environment are most important in deciding where to locate R&D, 65 percent of executives in the survey said the quality of the local education system is very or critically important. Related to this, proximity to major universities and research labs remains an important advantage for many types of R&D activity. When asked in which country the company was intending to spend the most R&D in the next 3 years, China was the most cited, the United States was number 2, and India number 3. The top 10 countries[6] show a mix of traditional and emerging centers of R&D.

R&D costs are escalating in high-tech industries, so it was not surprising that more than half the companies surveyed said R&D costs are an important benefit of a global R&D strategy and an important factor in determining a destination for R&D investments. According to the survey, companies weigh cost benefits in a range of areas—including lower-cost labor, cheaper land and office rental costs, and favorable tax regimes. However, it is noteworthy that cost is still less important than skills and expanding markets when it comes to determining a destination. The EIU survey also reports that companies consider the availability of local R&D expertise, among scientists and managers, as one of the most important factors. The EIU report suggests that one reason for this is that savings from cheaper labor are partially offset by the costs of coordinating R&D across multiple countries. How companies respond to these pressures often depends on size: Big companies are generally able to globalize R&D internally by opening their own

---

[5]EIU, *Scattering the Seeds of Invention: The Globalization of Research and Development* (2004), available at http://www.eiu.com/GlobalisationOfRandD. It is important to note that findings from such surveys are not scientific and may contain biases.

[6]The top 10 were China, United States, India, United Kingdom, Germany, Brazil, Japan, France, Italy, and Czech Republic.

BOX 2.2 Globalization and an American Company: The Timken Company

The Timken Company is a 105-year-old firm that has become a global company.[a] In 2003, it had more than $4 billion in sales spread over three primary businesses: metals (alloys and specialty steels), bearings for automotive applications (from engine to wheels), and bearings for industrial applications (from agriculture to aerospace). In recent years, Timken's primary application focus for its products has come to be power transmission and friction management.

When the company could not source steel of the purity needed for the bearings its customers wanted, it decided to develop its own steel. Timken's R&D has ever since mostly focused on the development aspect—that is, on developing solutions for very demanding applications like alloy tubing, gears, shafts, bearings, and so on.

Timken's R&D program also searches for new alloys, new thermal treatments, and new anticorrosion paths. Currently, the overwhelming majority of its R&D is carried out within the United States, including the R&D carried out under federal government contracts. However, occasionally some of the collaborations Timken engages in involve universities outside the United States. Nevertheless, essentially all the intellectual property has been and continues to be generated in the United States, although this may not continue indefinitely.

Timken is aware that investment in the more traditional subfields of materials research is decreasing in the United States and increasing in other parts of the world, like China. If this trend continues, Timken may need to consider doing R&D outside the United States in order to generate intellectual property for future products.

[a]This box is based on a teleconference between the panel's chair and Sal Miraglia, senior vice president of technology at the Timken Company, on October 20, 2004.

overseas laboratories, whereas medium-sized firms, constrained by cost considerations, may be more likely to globalize through outsourcing or alliances. Alternatives include the acquisition or licensing of existing technology in other countries (many companies do this by buying R&D expertise in other countries). Another increasingly important strategy in cost-sensitive industries is joint R&D ventures that enable companies to reduce substantially the time, cost, and risk involved in establishing overseas R&D operations.

Respondents to the EIU survey were also asked to name the biggest challenges of globalized R&D. Thirty-eight percent of the respondents said that robust pro-

Currently, most of the non-U.S.-based R&D sponsored by Timken is done to tailor oducts to local market needs. These limited excursions have demonstrated that there is a ge talent pool in places such as India and China. Accessing an equivalent talent pool in the nited States is hindered by deterrents such as unfavorable education trends, as well as ricter immigration controls and increasing security concerns, which make it more difficult employ noncitizens and recruit from abroad. On the educational side, the pool of univerties Timken hires from has shrunk to only a handful. Although advances in the kinds of search of interest to Timken (for example, alloy research) can be very challenging to hieve and therefore slow to emerge, they remain necessary for the health of the company.

While corporations such as Timken contemplate investments outside the United States, appears that opening technical and R&D centers abroad is only a matter of time. China's frastructure is growing, and incentives are being put in place to attract investment there. In dition, and perhaps most importantly, the market for products is burgeoning, and U.S. rporations such as Timken need to be close to their customers. The kinds of operations rporations contemplate vary from opening foreign subsidiaries to collaborating with local ppliers and manufacturers. But all these options only deepen the internationalization of rporate R&D and increase globalization.

There are, however, disincentives corporations consider when making decisions about ternational R&D investments. Maintaining the ownership of IP is a major concern, as are e economic and, at times, the human rights and political systems in the potential receiver untries. Indeed, protecting IP can trump economic concerns in such decisions.

U.S. companies can be expected to consider ethics to be very important for corporate olicy and to not do business where they cannot follow ethical practices. Furthermore, it ould not be expected that corporations like Timken will create IP in a country where the IP gime could compromise the company's global control of its IP. However, being close to stomers remains the number one reason for deciding where to locate R&D. The availabily of talent is another important reason.

tection for IP was critical and entered into their decision on where to base R&D more than any other business factor. Another 46 percent of respondents ranked it number 3 or 4 on the 5-point scale of importance used in the survey. Although China was ranked as the number 1 planned destination for new R&D, respondents to the survey expressed concern about the level of IP protection there. Along with IP concerns, 51 percent of the respondents said attracting top R&D talent was very important or critical, ranking it number 4 or 5. Other important challenges were identified: effective collaboration between international teams and compressing the time to commercialization.

BOX 2.3 An Evolving Model of Technology Transfer: The Transfer of Nanotechnology from New Zealand to an American Firm for Introduction to the Global Market

The globalization of the electronics industry, combined with the innovative and disruptive introduction of nanostructures in materials, is leading to new business models and flows of technology from research to the marketplace. The following shows how a start-up American firm with nanoprocessing technology and strong marketing links to major firms making electronics materials and their customers created a joint venture with a New Zealand firm to complement the U.S. firm's technology base and to help the New Zealand firm to transition technology from research into the marketplace.

NanoDynamics, a U.S. nanotechnology firm and manufacturer of nanomaterials, signed a joint venture agreement with a New Zealand-based technology company, Nano Cluster Devices Ltd. (NCD). NanoDynamics will be working to commercialize NCD's technology for the self-assembly of nanowires in the production of semiconductors and electronic components. Under the agreement, NanoDynamics will be responsible for sales and applications development, targeting semiconductor companies, consumer applications, and aerospace, biotech, and industrial manufacturers. NCD will be responsible for advancing the technology platform.

NanoDynamics is building an IP portfolio to provide significant product and technology value to a wide range of customers and partners. It is doing this through its own invention process, along with technology acquisitions and partnerships, extending its technology and product offerings to the markets it understands: the electronics, semiconductor, and energy markets. Its management team has launched commercial products from new technologies,[a] particularly in the area of advanced materials.

NCD, in conjunction with the University of Canterbury, developed a novel technology platform to produce electrically conducting nanowires by the deposition of atomic clusters onto lithographically prepared templates. The NCD technology produces small, well-controlled linear structures with different functionalities on a range of substrates. The key feature of its technology is that it is essentially a self-assembly process, which means that the slow manipulation of nanosized building blocks, which is often unavoidable in many other nanotechnologies, is completely avoided. The nanowires produced are smaller and more economical to apply due to their controlled placement and use of existing semiconductor processes.

From NCD's perspective the partnership with NanoDynamics is an important step in transferring its enabling technology to global industrial partners, to which NanoDynamics already offers advanced nanomaterial solutions.

[a]An example of such a product is NanoDynamics' new, patented golf ball, which combines state-of-the-art nanotech materials and construction. Details available at http://www.ndmxgolf.com/.

BOX 2.4 Recent Corporate Global Announcements

To reinforce its position as the leading coatings supplier in the growing Chinese automotive industry and accelerate the upgrading and expansion of production, DuPont has boosted its ownership stakes in a joint venture in automotive coatings businesses in China (March 2004). Agilent Technologies announced that it would increase its R&D workforce in India from 900 in 2004 to more than 2,500 by 2006, focusing on wireless solutions, billing software for telecom service providers, and device drivers for electronic products (July 2004). Cisco is set to build an R&D facility with 100 new hires over 18 months in China (September 2004). The company is joining top U.S. and European technology companies such as Motorola, Microsoft, IBM, SAP, and Oracle that have set up sales and manufacturing centers and R&D facilities in China. Cisco's CEO said that over the next decade, half of Cisco's top 12 business partners and half of its main competitors would come from China.

Nokia announced (July 2004) several R&D activity expansions in China, concentrating on Asian user interfaces; 3G and other radio technology; Internet Protocol; and Chinese mobile applications. Under this reorganization, 40 percent of Nokia's handsets will be designed and developed in Beijing. Motorola announced (July 2004) that it plans to merge its 19 R&D centers in China into one R&D firm. The company is expected to invest about $500 million over the next 4 years to strengthen Chinese R&D capabilities. Dell opened a center in China (September 2004) for multinational customers that need faster response times for IT services. Timken, a global leader in bearings and steel, has inaugurated an R&D center that will house 250 employees in Bangalore, India. It is Timken's largest R&D center outside the United States and the first in Asia. The center in Bangalore will support applications staff all over the world, design tools, do process engineering, and develop IT applications for the rest of the company.

GE has reorganized its R&D activities into GE Global Research, which is made up of 12 global laboratories organized by scientific discipline, all focused on leveraging technology breakthroughs across multiple GE businesses. GE Global Research consists of 2,500 employees working in four facilities: Niskayuna, New York; Bangalore, India (opened in September 2000); Shanghai, China (opened in October 2003); and Munich, Germany (opened in June 2004).

In addition to economic and market drivers for locating R&D in a particular foreign or domestic location, politics, both domestic and international, can be a driver. A study of political impacts and drivers is beyond the scope of this report, however.

One area of corporate MSE R&D that has undergone major changes in recent years is superalloy research. Its evolution is presented in Box 2.5, "Superalloy Case Study." A more detailed version of this case study can be found in Appendix F.

BOX 2.5 Superalloy Case Study

Superalloys are alloys based primarily on nickel and cobalt that have tremendously useful properties at elevated temperatures and/or in corrosive environments. The superalloy industry serves a limited market, albeit with critically important products. Superalloys find application in the aerospace industry as the enabling and primary material used in the hot end of jet engines—both rotating and static components—in the auxiliary power units used in aircraft, and in land-based industrial gas turbines. They are also used in petrochemical refining facilities where elevated temperatures are involved, in chemical plants where corrosive conditions can exist for which normal stainless steels are not useful, and in sour oil and gas wells. Maintaining a competitive position relies on managing operating costs and maintaining a competitive lead in state-of-the-art technology, which in turn requires continuing R&D on constantly evolving compositions and processing technology.

The number of U.S. superalloy R&D personnel decreased significantly in the past decade—by more than 50 percent in most companies and up to 100 percent in some. For example, Special Metals reduced R&D spending and personnel, from 50 or 60 engineers in the mid-1990s to about 5 at present. Allvac decreased R&D staffing to 14 from a peak of 30. Cartech has also cut back on R&D staff, from about 40 people in the 1990s to 20 today. From 1980 to 2004, the R&D effort at Haynes International decreased from 125 persons to 32 persons, and R&D spending was cut from $3.7 million in 2002 to an annualized rate of $2.4 million in 2004. International Nickel Co. (INCO) has gone from being the primary developer of nickel-base superalloy compositions to doing no R&D. The number of researchers at Howmet Research Center (Howmet is a wholly-owned subsidiary of Alcoa) has gone from 240 at the time of Alcoa's acquisition to 117 today. GE Engines has supplanted the R&D that was once done at the GE Corporate Research Lab and now works with partner Snecma (France) and suppliers of materials such as Cannon-Muskegon. It is also engaged in moving more investment overseas, particularly to China. Pratt & Whitney has cut back drastically on the number of researchers involved with superalloy R&D, concentrating on commercial engines and on solving supplier quality problems. Honeywell has cut back drastically on research and is essentially doing none. Solar Turbines has reduced alloy-development activities dramatically and relies almost exclusively on suppliers for new alloys and parts.

International activity is also evolving. Virtually all of the research to develop new superalloys in the United Kingdom is being carried out by primary producers and funded by Rolls-Royce. Germany's Krupp-VDM no longer has an R&D operation, and Asea Brown Boveri has cut back R&D in recent years. Superalloy research continues at government-supported laboratories such as the Max-Planck Institute, the Fraunhofer Laboratory, and so on. Italy's Acceria Foroni, regarded as a tough competitor in the marketplace, is investing in large-diameter (33-in) VAR technology with a view to producing 718 alloy ingots for rotating parts. Japan's Mitsubishi Materials and Daido Steel continue to be very active. Hitachi Heavy Industries has eliminated R&D at its laboratory but continues R&D in its plant. The main research institutes in China are the Central Iron and Steel Research Institute, the Beijing Institute of Aeronautic Materials, and the Institute of Metals Research. The Shanghai No. 5 steel plant is a large superalloy production facility with a large amount of the latest and best equipment in the world and a research group said to have 2,000 professionals.

The Special Metals Processing Consortium (SMPC), a U.S. consortium of production companies, has helped solve some difficult problems common to all which one company would have found difficult to solve alone. Originally there were 13 companies, but over the past 5 years the number of companies supporting SMPC has dwindled to six or seven, because several went into bankruptcy, suffered financial losses, or were merged out of existence. Government support has been obtained from Sandia National Laboratories (SNL) and the Federal Aviation Administration (FAA). However, SNL has indicated that the relevant laboratory there would be closed in 2005, and the amount (in real terms) of FAA research support has been cut by more than one-half over the last 15 years.

Universities that have conducted research on some aspect of physical metallurgy or processing of superalloys in recent times include the University of Michigan, Lehigh University, Purdue University, the University of Arizona, the University of Texas, Ohio State University, Penn State University, the University of Tennessee, Michigan Tech, the University of Pittsburgh, the University of Florida, and Northwestern University. NASA and the Air Force Research Laboratory are also active in superalloys research.

The superalloy industry has evolved to the point that production facilities are now divisions or subsidiaries of larger companies. Those that are still independent are widely thought to be facing difficult markets, with their customer base moving increasingly offshore. The main challenge for superalloy R&D is that it takes a long time to develop a new alloy, on the order of years, and to get customer acceptance and large orders takes even longer. Furthermore, new alloys often replace existing products, resulting in very low returns on R&D investments and little incentive to support the development of new alloys or processes. The lack of new products on the market is both a result of and a cause of the continued financial weakness of the industry over the past two decades.

Much alloy development is now being done in government-supported laboratories overseas. Process improvement continues to be done on a limited basis in the United States and is the key to commercial U.S. success for the time being. However, as new alloys are developed elsewhere and U.S. process know-how diffuses offshore, that competitive edge will disappear. It is possible that as the manufacturing end of the industry moves overseas, the ideas for research, which in the past came largely from working with customers, will move with it, because most alloy development is for the purpose of solving a customer problem or providing a new product idea. The result might be a drying up of research ideas in American companies.

continues

BOX 2.5 Continued

In conclusion, it is clear that U.S. superalloy R&D has declined significantly over the past decade. This is at least in part because U.S. firms that develop and manufacture superalloys face slower demand growth and higher costs, and many are in financial difficulties. Attracted by lower production costs and stronger growth in demand, superalloy manufacturers are increasingly looking to locate manufacturing overseas. U.S. companies that do so will probably stay competitive and survive, but only to the extent that they are privy to future developments in non-U.S. laboratories and plants.

SOME TRENDS IN MSE EDUCATION

An examination of globalization and MSE R&D would not be complete without some consideration of MSE education, even though a thorough review would be beyond the scope of this study, especially because MSE practitioners come from very diverse backgrounds, from engineering to natural sciences, making a comprehensive review very broad. Nevertheless, examining some of the issues related to education and globalization—for instance, trends in student enrollment and the supply of U.S.-trained practitioners and the role of non-U.S.-born researchers—provides an interesting perspective on the broader topics in this report.

Materials Education Today

If MSE professional and research positions are being lost in the United States in MSE subfields like superalloy research, then clearly there is a concomitant loss of skills, experience, and accumulated and tacit knowledge. This situation, in turn, can translate into lost R&D capacity and a loss in the ability to apply new ideas, products, and processes to national security and other national needs. How well is the country's MSE education system going to be able to react to and address this potential loss in capacity?

One immediately obvious challenge for MSE education, and indeed the education system in general, is the poor standing of U.S. high-school graduates in mathematics and science. According to the latest OECD triennial Program for International Student Assessment (PISA),[7] the United States ranks 20th in the

[7]PISA is an assessment (begun in 2000) that focuses on 15-year-olds' reading literacy, mathematics literacy, and science literacy. In the United States, this age corresponds largely to grade 9 and 10 students. PISA also measures general or cross-curricular competencies such as learning strategies. PISA emphasizes skills that students have acquired as they near the end of mandatory schooling.

TABLE 2.2 PISA (Program for International Student Assessment) Mean Scores in Science, 2003

Country	Mean Score	Country	Mean Score
Finland	548	Poland	498
Japan	548	Slovak Republic	495
Hong Kong (China)	539	Iceland	495
Korea	538	**United States**	**491**
Liechtenstein	525	Austria	491
Australia	525	Russian Federation	489
Macao (China)	525	Latvia	489
Netherlands	524	Spain	487
Czech Republic	523	Italy	486
New Zealand	521	Norway	484
Canada	519	Luxembourg	483
Switzerland	513	Greece	481
France	511	Denmark	475
Belgium	509	Portugal	468
Sweden	506	Turkey	434
Ireland	505	Mexico	405
Hungary	503	Brazil	390
Germany	502		

SOURCE: OECD, available at www.pisa.oecd.org.

world in the science literacy of its high school students (Table 2.2). Science literacy among freshman college students in MSE and other science and engineering fields is a challenge to universities in that they are faced with bringing the knowledge and skills of these students up to international standards. B.S. curricula, therefore, have to start with some course work in basic mathematics, chemistry, and physics.

Not only are students lacking in some basic skills in comparison with their international peers, but MSE education is also faced with low and—worse—declining enrollments of undergraduate students in materials courses. There are also interesting trends when enrollment in science and engineering (S&E) programs in the United States is compared with that in competitor countries.

Looking at data gathered by the OECD on global education trends,[8] in 2000 about 2.8 million degrees were in awarded around the globe in S&E fields[9]—more than 1 million in engineering, almost 850,000 in social and behavioral sciences, and almost 1 million in mathematics and natural, agricultural, and computer

[8]OECD, *Education Statistics and Indicators: Education at a Glance* (2002). Available at http://www.oecd.org/education.

[9]These worldwide totals include only countries for which data are readily available (primarily the Asian, European, and American regions) and are therefore an underestimation.

sciences combined. Asian universities accounted for almost 1.2 million of the world's S&E degrees in 2000, with almost 480,000 degrees in engineering. Students across Europe (including eastern Europe and Russia) earned more than 830,000 S&E degrees, and students in North America earned more than 500,000. The OECD reports that in 2000 the proportion of the college-age population who earned degrees in natural science and engineering was substantially larger in more than 16 countries in Asia and Europe than in the United States. The United States achieved a ratio of 5.7 per 100 after several decades of hovering between 4 and 5. Other countries and economies recorded bigger increases over similar time periods: South Korea and Taiwan increased their ratios from just over 2 per 100 in 1975 to 11 per 100 in 2000–2001. At the same time, several European countries doubled and tripled their ratios, reaching between 8 and 11 per 100.

In several emerging Asian countries and economies, the proportion of first university degrees earned in S&E was higher than in the United States. In 2001, 38.6 percent of degrees awarded in China were in engineering and 11.2 percent in the natural sciences. In South Korea in 2000 the numbers were 26.9 percent and 6.4 percent, respectively. In the European Union the numbers averaged 13.5 percent and 8.5 percent, respectively. In contrast, students in the United States in 2000 earned about 4.8 percent of their bachelor's degrees in engineering fields and about 6.6 percent in the natural sciences.

Figure 2.2 shows the number of degrees awarded in the United States in metallurgy and materials over the last few decades. It is noteworthy that almost the same number of B.S. degrees is granted today as in 1967, while over that time the U.S. population has doubled and college participation has increased significantly.[10] One consequence of the small total enrollment in MSE and a proliferation of programs—see Box 2.6, "Evolution of Materials Departments"—is low enrollment per program. For example, 18 MSE programs have fewer than 10 graduates per year and 25 have between 10 and 25 per year. Over 50 percent of the graduates each year come from just 15 programs.

With few students per class, MSE departments can find it extremely difficult to justify the cost and number of faculty having the expertise to cover MSE—performance, properties, synthesis and processing, and composition and microstructure—in any one program. The National Academy of Sciences report *Materials and Man's Needs* (1974) noted that there was a need for all engineers to receive more in the way of materials education. Arguably this statement is truer now than

[10]The National Center for Education Statistics reports that between 1967 and 2001 the percentage of all 18- to 24-year-old high school completers enrolling in degree-granting institutions increased from 33.7 percent to 44.2 percent. See *Digest of Educational Statistics* (2003), available at http://nces.ed.gov/programs/digest/d03/tables/dt188.asp.

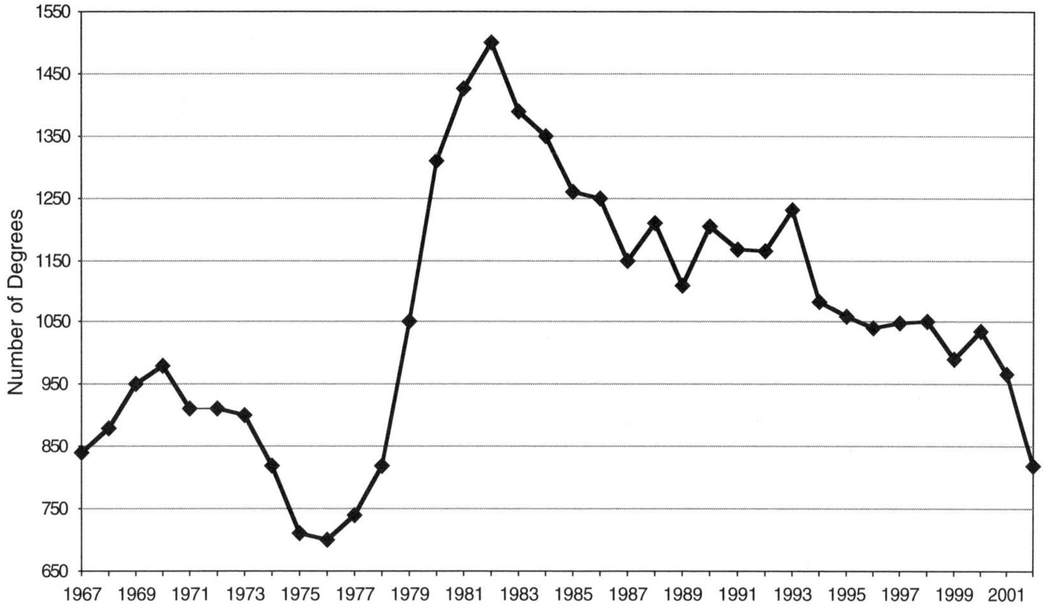

FIGURE 2.2 Number of B.S. degrees in metallurgy and materials science since 1967. The curve shows a steady decline since the early 1980s, a period during which the number of college-going students increased significantly. SOURCE: *Digest of Education Statistics*, 2003, available at http://nces.ed.gov/programs/digest/d03/tables/dt188.asp.

it was then, but there is no evidence that engineers other than materials engineers are receiving significantly more materials education today than in 1974. In fact, many of the other engineering fields have cut course work in materials from the curriculum or made it an elective. Rarely does a mechanical, civil, chemical, or electrical engineering graduate have more than a single survey course in materials. As an example, the number of manufacturing engineering programs grew from 2 in 1978 to 23 in 2003, but while some materials processing is taught in these programs, only a few have more than one survey course devoted to materials per se or more than two courses in materials processing.

The situation with graduate education is also challenging. Figure 2.3 indicates the steadily declining enrollment in MSE programs relative to enrollment in all engineering over the same period. Figure 2.4 indicates the number of graduate degrees in materials each year since 1990, with a decline over the most recent years. Some MSE practitioners say that many MSE graduate students come from a diverse S&E undergraduate background; sometimes this background includes MSE but sometimes it does not. So, they say, being in a graduate MSE program does not necessarily mean having a broad background in materials, and since graduate students are not often required to make up the MSE coursework they missed, the

BOX 2.6 Evolution of Materials Departments

In 1955 there were about 45 programs in the United States where metallurgy and metallurgical engineering were taught. The curricula, averaging 142 semester hours, covered not only metallic materials but also, by extension, semiconductors and other solid inorganic materials with interesting properties, including oxides. There were also 12 undergraduate programs in ceramic engineering but no undergraduate degree programs in polymer science or plastics. In 1959, DOD established three materials science centers at major universities. More were added in the next years, and some were then cut, with eight being transferred from DOD to NSF in the 1970s. This development caused former departments of metallurgy to morph into materials science departments in order to qualify for funding from DOD and the Advanced Research Projects Agency.

By 1974, more materials programs had emerged and 60 were ABET-accredited,[a] most offering at least 134 semester hours for the B.S. degree. The course work in these programs required engineering mechanics, physical chemistry, and electrical engineering. At that time, there were 12 ceramic engineering programs, and course work relating to polymeric materials and technology was largely taught in chemical engineering departments.

The watershed 1974 report *Materials and Man's Needs: Materials Science and Engineering*[b] sped up the transformation of materials programs, and the trend toward materials science took place despite the report's recognition of the need for what it termed curricular balance. The report recommended that, depending on local circumstances, materials-related degree programs should provide more emphasis on materials preparation and pro-

[a]ABET is the recognized accreditor for college and university programs in applied science, computing, engineering, and technology. See http://www.abet.org.

[b]National Academy of Sciences, *Materials and Man's Needs: Materials Science and Engineering*, Washington, D.C.: National Academy Press (1974).

depth of materials knowledge of some of those graduating with advanced MSE degrees comes into question.

Figure 2.5 shows that while the number of students graduating with a master's degree in materials is in decline, the number of foreign-born students graduating with that degree has held steady. This confirms the widespread belief that MSE has increasingly become dependent on attracting non-U.S. students. In many respects this is an early impact of the globalization of MSE R&D. In 2002, of the 286 doctorates granted in MSE in the United States, 151 were granted to foreign students—with Indian, Korean, and Chinese students being the most numerous. However, as these countries establish their own institutions of graduate education, prospective students will probably be less inclined to come to the United States.

cessing; polymer technology; design and systems analysis; computer modeling; and relations among the properties, function, and performance of materials. As programs continued to evolve, small-enrollment ceramic engineering departments merged with small-enrollment metallurgical engineering departments and became "materials" programs. This trend resulted in the disappearance of ceramic engineering departments. In 1977, the first undergraduate program in polymer science was accredited, and in 1978 the first program in plastics engineering was recognized. A common response from many materials programs was to include polymeric materials in their new curricula, so faculty were added and course work dealing with polymeric materials was developed. The greater emphasis on polymers resulted in a concomitant decrease in the time devoted to metals and ceramics. Coincidentally, many universities reduced the number semester hours required to obtain a B.S. degree to as few as 128.

Today there are 63 undergraduate materials/metallurgical programs, only 4 undergraduate degree programs in polymer science and engineering, and 6 undergraduate degree programs in ceramic engineering. Over the years, many MSE departments have broadened the scope of their curricula to cover all material classes. A criticism of this trend is that such a broad approach can risk cutting several corners of the field (such as processing or performance) and not covering some areas at all or in sufficient depth. For instance in a survey of materials curricula,[c] 50 percent neither required nor taught engineering mechanics and many required no course work in materials processing. Meanwhile, as discussed in this report elsewhere, undergraduate and graduate enrollment in MSE departments and schools is in decline.

[c]Curricula available on the Web were examined by the committee, and the trends are reported here.

Figure 2.6 shows the number of S&E Ph.D. degrees earned by Asian students within Asian and U.S. universities and compares it with the number awarded to Chinese students within Chinese and U.S. universities. There is no reason to expect that the trend in numbers shown in Figure 2.6 for S&E in general will be much different from the trend for MSE-degreed graduates in Asia, China, and the United States. A further problem for the field is that relatively few U.S.-produced MSE graduates enroll in graduate studies.

Given the dependence of MSE R&D in the United States on foreign talent, the data gathered by the NSF on graduated students who return to their home countries is also of interest. Historically, approximately 50 percent of foreign students who earned S&E degrees at universities in the United States planned to stay in the

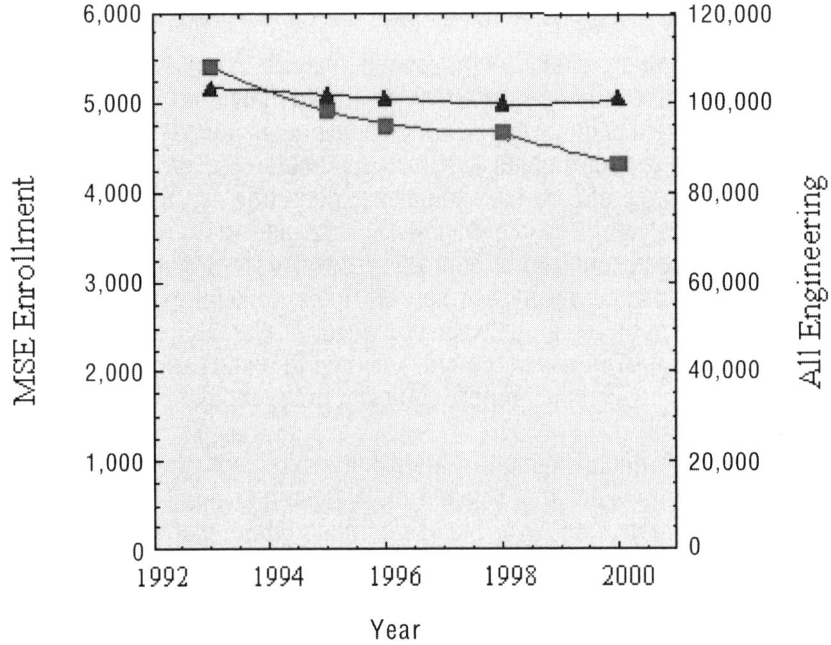

FIGURE 2.3 Enrollment in graduate MSE programs lags behind enrollment in engineering graduate programs in general. Graduate enrollment in engineering shows an 11 percent drop and in MSE a 21 percent drop for the period 1993–2000. Data from NSF, taken from presentation of Julia Weertman at the National Materials Advisory Board workshop Workforce and Education in Materials Science and Engineering: Is Action Needed? October 21, 2002, Irvine, California.

United States, and a smaller proportion said they had firm offers that would allow them to do so. These percentages increased significantly in the 1990s. From 1985 to 2000, most U.S. S&E doctoral degree recipients from China and India planned to remain in the United States for further study and employment. In 2001, 70 and 77 percent, respectively, reported accepting firm offers for employment or postdoctoral research in the United States. Recipients from South Korea and Taiwan are less likely to stay in the United States. Over the 1985–2000 period, only 26 percent of South Koreans and 31 percent of Taiwanese reported accepting firm offers to remain in the United States. Both the number of S&E students from these Asian economies and the number who intended to stay in the United States after receipt of their doctoral degree fell in the 1990s. This decline could be due to the efforts by Taiwan and South Korea to expand and improve their advanced S&E programs and create R&D institutions that offer more attractive careers for their

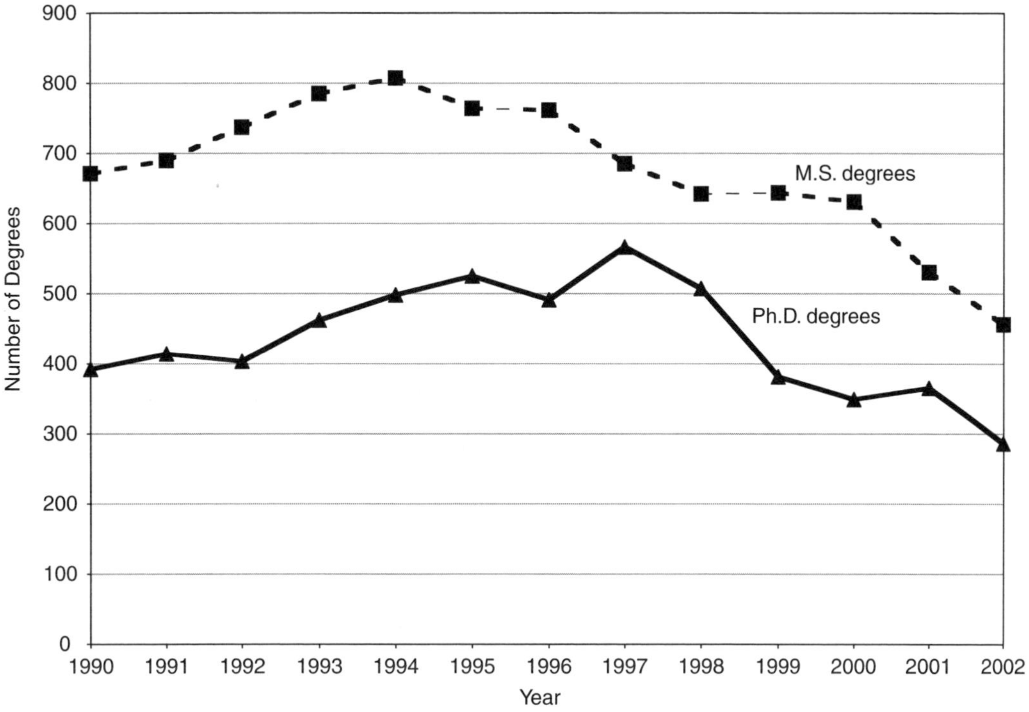

FIGURE 2.4 Graduate degrees awarded in metallurgy and materials engineering since 1990. The number of such degrees declined in recent years.

expatriate scientists and engineers. Notwithstanding these trends, by 2001 about 50 percent of their new U.S. doctorate holders reported accepting U.S. appointments.

Conclusion on Education

With the downturn in industrial activity in the MSE sector in the United States, the case can be made that the country does not need as many professionals in the materials field. However, the case can also be made that with improved productivity, manufacturing can be stabilized and, since materials are the enabling technology for most products, knowledge of materials is still very important to the future of the country. MSE professionals who can be productive on the domestic scene and competitive on the international scene must be well educated in all

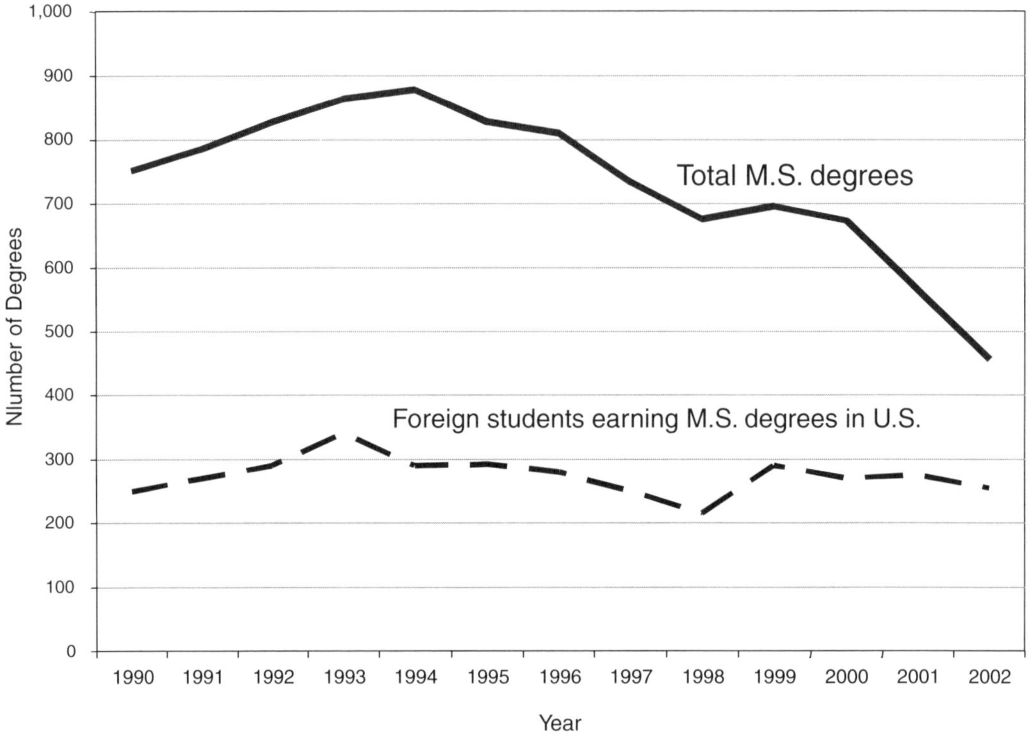

FIGURE 2.5 Master's degrees in materials: total and foreign students The number of foreign students graduating with a master's in materials has remained steady, while overall graduation numbers have declined. MSE graduate programs appear to be becoming more reliant on foreign students. SOURCE: American Society for Engineering Education.

aspects—performance, properties, synthesis and processing, and composition and microstructure—of a specific material, be it metallic, ceramic, electronic, or polymeric.

In summary, a number of challenges exist for the MSE educational system: the increasingly broad curricula in materials departments, the decreasing attraction of MSE as a career choice for high school and university graduates, and the continuing dependence of graduate programs on attracting foreign students in an increasingly competitive global market for the best students. It is not clear that the current MSE education system, including research at universities, can meet the nation's needs by producing graduates with the required depth of knowledge.

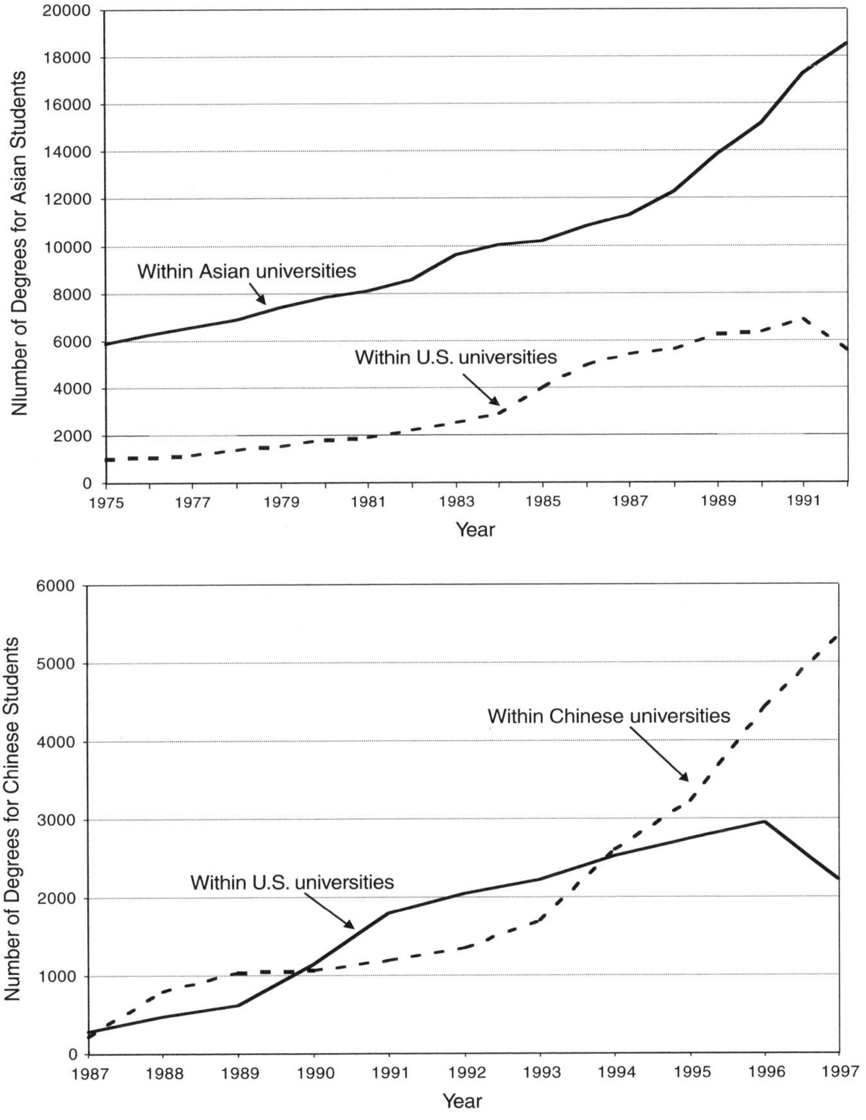

FIGURE 2.6 Comparison of S&E doctoral degrees earned by Asian students at Asian and U.S. universities (upper) and by Chinese students at Chinese and U.S. universities (lower). Data in the upper figure include degrees earned at universities in selected Asian countries: China, India, Japan, South Korea, and Taiwan. Asian students at U.S. universities include students on either temporary or permanent visas from China, Hong Kong, India, Japan, South Korea, Taiwan, and Thailand. In both cases the number of students completing their doctoral science and engineering degrees in Asian/ Chinese universities is increasing rapidly while fewer Asians are attending U.S. universities following continuing growth in earlier years. SOURCE: NSF, *Science and Engineering Indicators* (2000).

Conclusion. The MSE education system, including K–12 mathematics and science education, will have to evolve and adapt so as to ensure a supply of MSE professionals educated to meet U.S. national needs for MSE expertise and to compete on the global MSE R&D stage. The evolution of the U.S. education system will have to take into account the materials needs identified by the federal agencies that support MSE R&D as well the needs of the materials industry.

SUMMARY REMARKS

In summary, the evidence presented in this chapter and its associated appendixes—patent data, literature data, trends in corporate research—indicates increasing activity in MSE R&D around the world, with concomitant increases in global and transnational ownership and collaboration. In the subfields of MSE, the United States is the leader or among the leaders, and Japan and Western Europe are the closest rivals for leadership. Global activity in all the subfields examined is diversifying, with significant increases in activity in Asian countries that have not heretofore had substantive activity in these fields. On the home front, MSE R&D in the United States appears to be a highly internationalized activity with a highly diversified set of international partners. Advances in IT and communications technology have been drivers for the globalization of R&D. The case study in superalloys (Appendix D) shows that the environment for R&D in this field is evolving in the face of globalization. A deeper analysis of the current state of R&D in 10 MSE subfields is presented in Chapter 3.

3

Benchmarking of Materials Science and Engineering R&D

This chapter reviews the outcomes of an earlier National Academies materials benchmarking study and considers the current status with a snapshot assessment of the status of each materials subfield.

In 1993, the Committee on Science, Engineering, and Public Policy (COSEPUP) of the National Academy of Sciences, the National Academy of Engineering, and the Institute of Medicine issued the report *Science, Technology, and the Federal Government: National Goals for a New Era* (the Goals report). In that report, COSEPUP recommended that the United States aim to be among the world leaders in all major fields of science so that it could quickly apply and extend advances in science wherever they occur. In addition, the report recommended that the United States maintain clear leadership in fields that are tied to national objectives, that capture the imagination of society, or that have a multiplicative effect on other scientific advances. To measure international leadership, COSEPUP recommended the establishment of independent panels that would conduct comparative international assessments of the scientific accomplishments in particular research fields. COSEPUP believed that these panels should consist of researchers who worked in the specific fields under review (both in the United States and abroad), people who worked in closely related fields, and research users who followed the fields closely.

In response, three panels were established to conduct evaluations of mathematics, materials science and engineering, and immunology. Each panel wrote its own individual report.[1] The individual reports were submitted to COSEPUP and published in a single report with an overview by COSEPUP.[2] The resulting benchmarking study on materials science is of particular interest to this report, which requires an understanding of the status of MSE R&D in the United States relative to the rest of the world.[3]

THE 2000 BENCHMARKING REPORT

In 1997 COSEPUP organized a benchmarking experiment that resulted in the 2000 benchmarking report. The experiment was carried out for three fields—mathematics, materials, and immunology—to learn if research leadership in a field could be assessed in a timely fashion at a reasonable cost. It examined the position of U.S. research in the selected field relative to research being carried out in other regions/countries; predicted relative U.S. future status based on the observed trends; and identified the key determinants of U.S. performance in the fields. Assessment tools for the studies in all three research fields were these:

- A "virtual" congress, with each panel asked to identify key invitees to a hypothetical international congress convened to address five or six hot topics. The aim was to identify the best of the best researchers from around the world and then use the information to construct tables ranking countries by the number of nominated invitees;
- Citation and journal analysis;
- Quantitative data such as the number of graduate students and subfield funding;
- Prize analysis; and
- Analysis of actual international congress speakers.

---

[1]National Academy of Sciences, National Academy of Engineering, Institute of Medicine, *International Benchmarking of U.S. Mathematics Research*, Washington, D.C.: National Academy Press (1997); National Academy of Sciences, National Academy of Engineering, Institute of Medicine, *International Benchmarking of U.S. Materials Science and Engineering Research*, Washington, D.C.: National Academy Press (1998); National Academy of Sciences, National Academy of Engineering, Institute of Medicine, *International Benchmarking of US Immunology Research*, Washington, D.C.: National Academy Press (1999).

[2]National Academy of Sciences, National Academy of Engineering, Institute of Medicine, *Experiments in International Benchmarking of U.S. Research Fields*, Committee on Science, Engineering, and Public Policy, Washington, D.C.: National Academy Press (2000); referred to hereinafter as the "2000 benchmarking report."

[3]It was not within the scope of this study to repeat the complete benchmarking process recommended and executed in the 2000 benchmarking report.

In considering the likely future U.S. position in research relative to other countries, the panels considered the intellectual quality of its researchers and its ability to attract talented researchers; its ability to strengthen interdisciplinary research; and its ability to maintain strong, research-based graduate education and a strong technological infrastructure. The panels considered cooperation among the government, industrial, and academic sectors and increased competition from European and other countries. They also looked at the evolution of federal support for research.

The report of the materials benchmarking panel[4] addressed the following subfields: biomaterials, composites, magnetic materials, metals, electronic and optical-photonic materials, superconducting materials, polymers, and catalysts. The panel determined that the United States was among the world leaders in all the subfields of materials science and engineering research and the leader in some subfields, although not in the field as a whole. An area of U.S. weakness in most subfields was materials synthesis and processing. Increasingly, the panel said, U.S. researchers were relying on specialty materials suppliers in Europe and Japan for bulk crystals and other specialty materials. The United States was identified as the "clear leader" in biomaterials and as the "leader" in metals and electronic-photonic materials. However, the lead in electronic-photonic materials was characterized as "endangered" because of cutbacks in industrial exploratory research. The panel concluded while the United States had once been preeminent in magnetic materials, at the time of the benchmarking it was "one of several leaders" in that subfield. In addition, the panel warned that U.S. leadership in composites, catalysts, polymers, and biomaterials was likely to be eroded because of the high priorities given to these subfields by other countries.

The panel concluded that the health of MSE R&D was dependent on researchers coming from other countries and that there was, accordingly, an associated risk of those countries building their own infrastructures and becoming more attractive as locations for internationally mobile researchers. The panel also concluded that a key determinant of U.S. leadership in MSE was its innovation system—that is, the entrepreneurial abilities of its researchers and the influence of its diverse economy. In addition, the panel said that the nation was strong in MSE by virtue of its intellectual and human diversity and its ability to draw intellectually from the entire science and engineering research infrastructure. The benchmarking panel warned that the ability of the United States to capitalize on its leadership opportunities could be undermined by shifting federal and industry priorities, a potential reduction in access to foreign talent, and deteriorating facilities for materials char-

[4]The materials benchmarking report was released in 1998 and was included as an attachment in the 2000 benchmarking report.

acterization. Of particular concern was inadequate funding to modernize major research facilities in the United States and to plan and build the new facilities needed to maintain research leadership.

The panel wrote that the U.S. position, "among the world leaders," was likely to slip in some areas of MSE. The reasons varied but included the globalization of research and the growth of other economies. The panel predicted that the United States could be expected to continue supporting materials science and picked out, among others, nanostructured materials and intelligent materials as exciting emerging areas of study. In addition, the panel opined that the United States would "never want to lose its current strength in aerospace and defense, which are important not only to U.S. security but for the stability of the post-cold-war world."

SNAPSHOTS OF THE CURRENT STATUS OF MATERIALS SUBFIELDS

As well as experimenting with the benchmarking process itself and analyzing the then-current status of U.S. research in each of the materials subfields, the 2000 benchmarking report pondered possible future developments for each of the subfields. To assess the impacts of today's level of globalization on MSE R&D, the predictions of that report are presented herein alongside a current snapshot of benchmarking analysis for each of the subfields.[5] The predictions from the 2000 benchmarking report are summarized in the bulleted introduction to each section.[6]

Biomaterials

- The 2000 benchmarking report stated that U.S. strength relative to other countries in basic and applied biomaterials research was likely to erode in the near and longer term for several reasons, including European and Asian (Japan, Singapore) investment, the cost and complexity of obtaining from

[5]Note that nanomaterials was not treated as a separate subfield in the 2000 benchmarking report but is so treated in this report.

[6]The benchmarking process undertaken for the present study was not as thorough as that for the 2000 benchmarking report, which would have been beyond the scope of this study. The idea here was to take a snapshot of the current status of a subfield, as provided to the committee by experts in that subfield. The snapshots focused on assessing where the subfield is relative to the predictions made in the 2000 report. Obviously there are limitations and uncertainties in the assessment of relative positions and capabilities using this or even the more thorough process used for the 2000 benchmarking report.

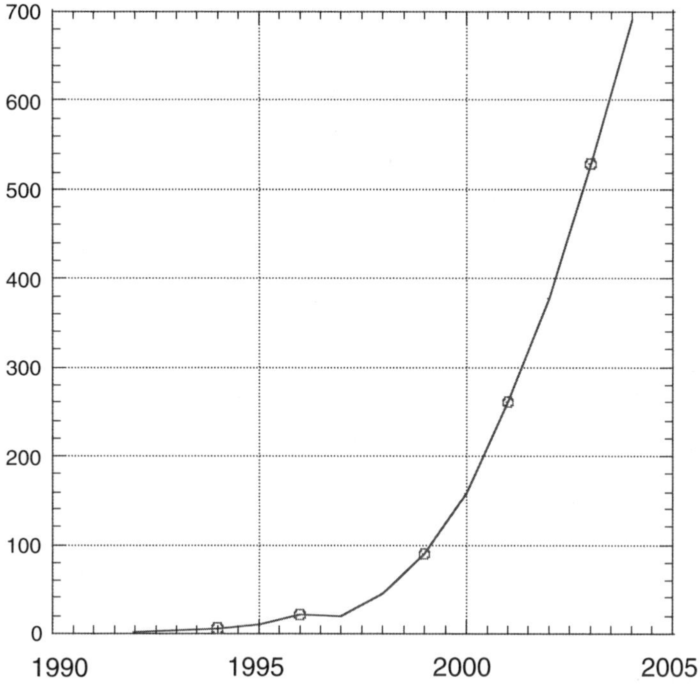

FIGURE 3.1 Total number of papers published in 1994, 1996, 1998, 2001, and 2003 with "biomaterials" in the title, worldwide. Of these, close to 60 percent originated in the United States. SOURCE: Scifinder Scholar database.

the Food and Drug Administration (FDA) approval for new biomaterials for in vivo use, and the litigious climate in the United States.

Biomaterials[7] research in the United States has grown over the past half decade as it has worldwide, and the United States is still in a leadership position although there are increasingly strong and focused efforts going on throughout the world. Much of the growth in biomaterials research in the United States was catalyzed by rapid growth in the number and size of U.S. biomedical engineering departments, catalyzed by a focused infusion of money from federal, state, and private sources. One indicator of this growth is the large increase in the number of biomaterials publications in the later 20th and early 21st century, as shown in Figure 3.1.

[7]For the purposes of this report, biomaterials are defined as materials developed for in vivo application.

While the focus on biomaterials research in academia is increasing, the nature of the research is changing. By the end of the 1970s many common materials, including polyethylene, poly(ethylene terephthalate), and several stainless steel and titanium alloys were approved for in vivo use by the FDA. Devices based on these materials have served the medical community well but have finite—on average, years to more than a decade—lifetimes. The materials, for the most part nonresorbable in nature, are the mainstay of the medical device industry. The difficulty and cost of bringing new materials through the FDA has led to a focus on modifying already approved materials for specific biomedical applications rather than introducing new materials, so that, since the 1970s, few new biomaterials have been commercialized. The complexity, in particular with respect to regulatory and other nonscience aspects, of introducing a new material into the marketplace has shifted clinical evaluation of new materials or devices to Asia or Europe. The reality, though, is that the United States is the largest market for medical devices in the world, so no matter where the initial research and clinical verification might be done, products with significant market potential ultimately are qualified for use in the United States, no matter how costly or difficult the process. The net result has been a rapid globalizing of biomaterials research rather than a diminishment of the U.S. effort.

Also contributing to the growth of U.S. biomaterials research is the commercial success of several materials-enabled devices, most notably drug-eluding stents and replacement spinal disks. In both devices materials must be matched to function, so biomaterials expertise is a fundamental need in the R&D teams bringing the device to commercial fruition. Another factor contributing to growth is that the materials being used are limiting the performance of many biomedical devices—for example, short-term degradation of electrical signals from implanted devices as a result of the body's reaction to implanted electrodes and conductors. The burgeoning of interest in tissue engineering and stem cells is stimulating the study of biomaterials as a source of improved scaffolds for cell growth and differentiation. Finally, there is increasing focus on materials for biosensors and biochips and on how the study of materials can complement the study of cell biology and molecular biology.

While the quantity of biomaterials research is large and growing, most of the work is highly observational. Little is understood about cellular or tissue response to material structure, from the nano through the macro, and there is a debate between the materials and biology communities about whether the material per se is a factor. Most existing biologically relevant materials, from metals through polymers, were approved for in vivo use decades ago, and modern research examines how surface chemical modification and gross device architecture affects efficacy. There is an opportunity to apply the classical principles of materials science to the biomaterials area with the expectation of both scientific and clinical success. The science of biomaterials is changing. What was once the dominant challenge—the development of new materials—has given way to a new one: the creation of an optimized interface between synthetic and biological systems. There is no evidence that U.S. leadership is threatened in any of these areas.

Ceramics

- The 2000 benchmarking report noted that, with the exception of research in electromechanical systems and coatings, the relative U.S. position in ceramics was in decline. Japan and Germany were predicted to continue to be highly competitive in the engineering of ceramics. The report also projected a potential for U.S. leadership in areas concerned with functional electronic ceramics, such as self-assembly materials and multilayer ferrite processing. Areas where the United States was among the world leaders and where the United States should maintain position in the future included three-dimensional nanoporous silicates, microwave dielectrics, and electrophoretic preparation of thin films. The United States was not expected to seriously challenge the Japanese leadership position in integrated micromagnetics.

Since the last benchmarking of ceramics R&D, major changes in the profession and the domestic industry have continued. The use of ceramic materials continues to grow, and they remain critical components of many larger materials systems and commercial products. The field of ceramics has continued to expand and is providing new opportunities and challenging the traditional view of technical ceramics. For example, work on ceramic materials for electronic devices has significantly expanded into both thin films and nanostructured materials, and the need for higher performance wireless systems has pushed the ceramics field to develop new microwave materials. Similarly, an earlier focus on passive optics is being replaced by a focus on active optics, such as optical amplifiers, and efforts in bioceramics and in ceramics for power sources, sensors, filters, and the like are rapidly expanding.

Ceramic materials for electronic substrates and for use in high-wear applications remain key commercial applications. However, since the 2000 benchmarking report, the primary areas of commercial and R&D growth for ceramics have been ceramic armor, fuel cells, and nanostructured materials. Ceramic armor is nearly as tough as more traditional metals with only a fraction of the weight. This property makes ceramic materials attractive for both personal and vehicular armor. Fuel-cell research, a priority of the government's energy program, is driven by the high cost and uncertain supply of fossil fuels. Ceramics also play a critical role in the commercialization of the government's hydrogen energy initiative.[8] Nanostructured ceramic materials are beginning to find a wide array of industrial, electronic, and biological uses. Other prominent or emerging areas in ceramics R&D include ultra-high-temperature ceramics for hypersonics; diamond and diamond-like films; ferroelectrics and piezoelectrics; ceramics with structured porosity; rapid prototyping and direct write processing techniques; and nonlinear optics.

The late 1980s and early/mid-1990s are viewed by some as having been the pinnacle of U.S. R&D in ceramics, with the focus on high-temperature superconductivity research and the emergence of large government programs on advanced structural ceramics. Since the late 1990s, the domestic industry has seen continued consolidation and dramatically increased competition from China and other Asian nations. In the 2000 benchmarking report, Japan was noted as "sharing the R&D leadership" role with the United States, and Japan remains a strong force in ceramics R&D. However, since the 2000 report, Korea has emerged as an important player in ceramics R&D and China is beginning to emerge as one. Western Europe, notably Germany, continues to play a major role in R&D for ceramics. It appears the U.S. share of worldwide ceramics R&D is decreasing. Many traditional ceramic areas like small electrical insulators and other lightweight, easy-to-ship materials have been especially affected. Additionally, the majority of the traditional university programs in ceramic engineering and science have merged into larger materials science and engineering programs.

While the United States has active programs for the study of ceramic materials, international competition is strong and becoming stronger. Even though U.S.-based industries continue to conduct the majority of their R&D activities within the United States, there is evidence that ceramics R&D, like manufacturing, is becoming a global enterprise. For example, East European countries and Russia have emerged as leaders in research on nonoxide glasses, with portions of the R&D activity being funded by the local governments, but U.S.-based industry has also recognized the expertise and has supported some of the R&D there. With other nations likely to continue to grow their domestic ceramics programs, foreign expertise in ceramics will strengthen.

Composites

- Basic research into composites at U.S. universities was identified in the 2000 benchmarking report as having come to a standstill as a result of the

[8]See, for example, the Department of Energy (DOE) report *Basic Research Needs for the H_2 Economy* (2004). Available at http://www.sc.doe.gov/BES.

> Department of Defense decision to strictly curtail university research funding in metal, polymer, and ceramic matrix composites. The panel warned that if that situation persisted, the United States would forfeit its leadership role in composites.

As recognized in the 2000 benchmarking report, U.S. leadership in composites was once unchallenged. Composites as a concept were actively pursued under Project Forecast, a 1963 DOD initiative that identified a number of technologies of such importance that sustained and significant funding was warranted. The 1960s through the 1980s, which saw substantial funding and activity in the subfield, resulted in the development of composites for projects such as the F-22 fighter jet and orbital satellite systems that are currently being spun off into systems for infrastructure, offshore oil, wind energy, transportation, and so on. Since then, the leadership position has been challenged—not least of all because of European investments in the commercial and military aerospace industries.

Europe has moved to the front line of composites manufacturing and modeling, particularly in polymeric composites. The decline of the U.S. ballistic missile program and reentry systems has led to a decline in carbon-carbon composites programs in the United States. Other countries, such as France, Korea, Taiwan, China, and Japan, continue to have active programs. While U.S. professional societies[9] and standards organizations[10] in the subfield continue to be prominent, their dominance is being challenged by their European counterparts. It is interesting to note that today the largest composites trade show in the world is held in Paris, not in the United States.

Composites may be the first step toward the next generation of materials, components, and structures, and continued understanding of their potential will result in significant system-level payoffs for multiple applications. However, to date, developments in composites have only scratched the surface of their potential, with acceptance having been achieved for only the very basic laminates and materials solutions still not designed to take advantage of the anisotropic and tailorable nature of composites. Hence there are many potential growth areas.

Nanotailored composites are one area of promise. Since composites are all about the joining together of dissimilar constituents and interfaces, the ability to characterize and engineer at the nanoscale will open up a new suite of multifunc-

---

[9]U.S. professional societies include the American Institute of Aeronautics and Astronautics (AIAA), the Society for the Advancement of Materials and Process Engineering (SAMPE), and the American Society of Mechanical Engineers (ASME).

[10]U.S. standards organizations include the American Society for Testing and Materials (ASTM) and the MIL-17 composite materials handbook organization.

tional composites for rapid insertion. The tailorability of composites also holds great promise. There is potential for tailoring multiple constituents for multifunctional response in three-dimensional space; tailoring through hybrid constituents; tailoring through bio-inspiration; and tailoring through asymmetric architectures and gradient morphologies. Accurate analysis capabilities will help predict material behavior at a level that allows innovative material design and engineering such as three-dimensional, complex, and hybrid composites (as opposed to simple laminates). Such tailorability could increase damage tolerance and reduce processing, joining, and fastening costs. The ability to accurately understand and predict composite behavior in complex environments will help move beyond the current practice of applying conservative safety factors and relying on empiricism. There is also great potential for multifunctionality, including designing the composite to perform structurally as well as handle other specific tasks—for example, thermal management and electrical management; monitoring and sensing; or providing shape morphing.

Without a long-term investment in composites research in the United States, the industry will stagnate and eventually become uncompetitive with foreign companies that have maintained active research programs. The economic impact for commercial aerospace products can be great. For example, the Boeing 787 (Dreamliner) has a clear opportunity to dominate Airbus products primarily owing to its large percentage of composites, which results in lighter weight and allows it to fly more passengers, and increased passenger comfort (higher cabin humidity), enabled by the corrosion resistance of the composites. Japanese fibers and prepregs are well known to have tighter distributions of properties, which increases design allowables and improves manufacturing.

The decline of U.S. leadership in composites research has been accompanied by fewer government- and industry-funded programs and an erosion of the academic base in the subfield. Key researchers are thought to be leaving the area and going into other fields where research is better supported. Program changes in federal agencies that support composites research, such as NASA and parts of the Navy, along with tight budgets, have led to a decrease in U.S. activity. Industry has pulled much funding away from R&D for new materials technologies.[11] Existing platforms and programs suffer from being risk-averse—that is, from wanting to use existing materials—and manufacturers do not want to disrupt production to insert new materials. This places a cap on development.

Today there are fewer U.S. commercial carbon-carbon manufacturers and far fewer companies providing oxidation coatings than there were 10 years ago. As a

---

[11]International SAMPE Technical Conference, *Report of the Panel on Transition* (2003).

result, there is little work ongoing to address sustainment and replacement of the shuttle fleet's carbon-carbon leading edge or next-generation carbon-carbon components for space or air applications. Today, the large commercial and defense programs are outsourcing production and supporting R&D overseas—for instance, in the Joint Strike Fighter program and the Boeing Company's Dreamliner program. In the United States, there is little work on computational or experimental benchmarks to accelerate insertion/certification.

In summary, the evidence presented here, along with the literature and patent data elsewhere in this report, suggests that the United States risks being unable to exploit the promise of composites because of the significant and continuing decline of its leadership in the subfield.

Magnetic Materials

- In the 2000 benchmarking report, the United States was described as "catching up with the leaders in international research on magnetic materials and magnetism." The panel found that the vitality of magnetic recording and the phenomenon of colossal magnetoresistance were starting to produce a renaissance in fundamental magnetism research in the United States.

The 2000 report did not address some areas in magnetic materials research that have become important to the field in the last 5 years. These include research on "hard" ferromagnets for more efficient, smaller, and lighter electric generators and motors; research on "soft" ferromagnets to decrease coercivities and for use in more efficient transformers that consume less energy; research on ferrites for improved high-frequency operation (especially for radar); and research on magneto-optic materials for information storage. More recently, there has been an increase in research into biomagnetism and materials for magnetic refrigeration. The biomagnetism area includes targeted drug delivery; detection and separation of antigens; magnetically improved magnetic resonance imaging (MRI) resolution in localized areas; localized control of biological or cellular activity; and magnetic heating probes for local thermal treatment.

Hard ferromagnet activity is centered in the United States and Europe (Germany), while Japan and Europe are the main centers of activity in soft ferromagnets. Centers of excellence in biomagnetism are found in the United States, Canada, Australia, and Europe, and magnetic cooling work is going on in the United States, Canada, Japan, Europe, Russia, China, and Hong Kong.

The United States is leading in magnetic refrigeration materials, with Europe (Netherlands) being a close second and the Pacific Rim in third place. This area of research is becoming such a hot area that a special session at the Magnetism and

Magnetic Materials Conference in November 2004 was devoted to it. Research in this area promises to find alternative cooling technologies.

In hard ferromagnets, Europe (Germany, in particular) and the United States share the leadership in recent research. Since the United States lacks the raw elements (rare earth elements, in particular) for permanent magnets, its permanent magnet industry is almost nonexistent. U.S. leadership in this area derives from the large number of users of permanent magnets here and their search for better materials in order to reduce the cost and weight of existing permanent magnets. In soft ferromagnets, Japan has the lead, with Europe (Germany) and the United States following. The Japanese have been particularly industrious at developing nanocomposites from rapidly solidified materials to create very low coercivity materials, especially for higher frequency operation. There are no U.S. manufacturers of soft ferromagnetic materials, but there are lots of U.S. users.

The United States is leading in magnetostriction materials, which are important mainly because of their application to microelectromechanical systems (MEMS) devices. The materials of most recent interest are FeGa alloys and alloys of GdGeSi, both of which have been found to possess large magnetostriction effects. Germany is quickly incorporating magnetostrictive materials into devices, including MEMS devices, and probably leads in that area. Japan is leading in research in magneto-optic materials for information storage, with the United States in second place, and the United States and Japan are probably tied for leadership in the area of giant magnetoresistance (GMR) devices, with Europe being a close second. However, with the sale by IBM of its magnetic recording activities to Hitachi, the only remaining U.S. manufacturer of magnetic recording devices is Seagate Corp. Leadership in this field of research will mean continued access to the large magnetic recording market.

Leadership in biomagnetism remains unclear as the field continues to grow in importance. It is also too early to tell how much impact magnetism will have in this area. Magnetism presently plays a strong role in the separation of biological species from a solution so they can be quantitatively measured. In addition, MRI resolution can be improved by placing magnetic particles in the vicinity of the area being imaged. How useful magnetism will be in drug delivery (another area with potential) remains to be seen.

In summary, leadership in the subfield of magnetic materials is mixed, with the United States in the lead in some critical areas and among the leaders in others.

Metals

- The 2000 benchmarking report concluded that "in all probability, the U.S. lead will remain, but that is not a certainty." The panel concluded that the

> United States' position as the only remaining superpower would be the driving force behind much MSE national-security-based research. However, the panel believed that with the end of the cold war, this pressure would diminish in proportion to perceived threats to national security, shifting the burden for materials development to nondefense industries. Another force that was identified as likely to affect the U.S. position was the consolidation and globalization of industries from aerospace companies to automotive suppliers. "For these businesses, the issues of U.S. competitiveness and research and development leadership are much less important, because their playing field is the world and they will seek knowledge wherever it is to be found."

Research into the production, processing, and development of metallic materials in the United States has continued to decline since 1998. Very little alloy development is being done by metal producers, which formerly did most of this work, and companies in the metal-consuming industries have also decreased their efforts. The very substantial decline in activity over 20 years or so is illustrated by the data in Table 3.1.

A continuing loss of trained personnel in metallurgy results in a concomitant loss of corporate expertise in producing, refining, processing, and applying the metals that remain the basic engineered material of industry. This problem is exacerbated by a decreasing emphasis in the education of "materials" students on the structure, properties, processing, and application of metallic materials, leading to concerns that there will be a shortage of well-prepared graduating students.

As productivity in industries like the steel industry and the foundry industry improved greatly, the relatively flat annual demand for steel and castings over the past two decades caused employment to decline. Employment opportunities in R&D have also declined, resulting in a significant decrease in the number of students pursing careers in metals R&D.

The 2000 benchmarking report stated that computer modeling of material processing was the strength of the U.S. industry. Indeed, some industries today are utilizing computer-based models of solidification and mechanical working, but it is not true that the United States is ahead of the rest of the world in this area. Developers and researchers in Japan and Europe have provided many of the models used in the metals industry for process modeling and control. A majority of presentations at the 2004 Materials Science and Technology Conference, sponsored by The Minerals, Metals and Materials Society (primarily a metallurgical society) and by the Association of Iron and Steel Technology (primarily steel makers), were by non-U.S. participants. This is another indication of the continuing decline in metals R&D in the United States. In summary, the evidence pre-

TABLE 3.1 The Loss of Metals-Related R&D Jobs in the United States Since 1980[a]

Research Laboratories	No. of Employees 1980	Year Closed	2004
Nonferrous industry			
Anaconda	100	1983	0
ASARCO	95	1992	0
St. Joe Zinc	100	1988	0
New Jersey Zinc	25	1980	0
Phelps Dodge	75		80
Kennecott Copper (Utah)	170	1990	0
Kennecott Ledgemont Lab	250	1982	0
Alcoa (St. Louis)	200		0
Alcoa (Pittsburgh)	~1,400		~400
Kaiser Aluminum	575	2000	0
Reynolds Aluminum	250	1999	0
Duval Copper	90	1983	0
Foote Mineral	50	1983	0
UOP	50	1983	0
Steel industry			
United States Steel	1,900		100
Bethlehem Steel	1,000	1982	0
Jones and Laughlin (Pittsburgh)	300	1984	0
Inland Steel (Ispat now)	300		100
Republic Steel (LTV)	200	2001	0
Allegheny-Ludlum	250		20
Youngstown Sheet & Tube	150	1970	0
Carpenter	75		20
Crucible Steel	150		10
Armco	150		20
Cleveland-Cliffs	320		32
Hanna Mining	75		0
Alloy industry			
International Nickel (New York)	300	1980	0
International Nickel (West Virginia)	100		20
Union Carbide (Tonawanda)	300	1988	0
Amax (Golden+Ann Arbor)	300	1986	0
Special Metals	20		5
Haynes Stellite	100		20
Cabot Corporation	150		10
Allvac	10		30
Independent metals research labs			
Battelle Columbus (metals)	150		5
IITRI (metals)	100		5
Hazen Research	180		75
U.S. Bureau of Mines	1,800		50
Colorado School of Mines Research Institute	200	1987	0
Mountain States	50		3

[a]Data in this table were gathered from one-on-one interviews conducted by the committee.

sented here and the literature and patent data reported in Chapter 2 suggest the United States appears to be losing its leadership, and there are no indications that this trend is going to be reversed any time soon.

Electronic and Optical-Photonic Materials

- The 2000 benchmarking report predicted that research in electronics would continue to focus on materials and processes and that it would be conducted globally through international collaborations among industrial organizations. Industry partnerships with academia and government, along with focused centers at universities, were expected to continue to be vitally important to U.S. leadership in the now-global semiconductor industry. The panel predicted that as semiconductor technology approached the 100 nm scale, the United States and others would make these advances more or less equally, if not as partners. The panel concluded that the United States would continue its leadership position in compound semiconductors (GaAs, GaAlAs) and wide-band-gap semiconductors (SiC) for power devices and microwave transmitters. Europe was expected to continue to share leadership with the United States in electrical power distribution and motor control applications of power transistors. In the field of nanotechnology, the panel noted that the United States had traditionally been the leader in exploratory nanostructures, including quantum wires and dots, and that it shared the lead with Europe and Russia in mesoscopic physics. However, the United States had conceded commercial leadership in wide-band-gap photonics to Europe and Japan, and the Japanese enjoyed "a commanding lead" in GaN technology and the commercialization of photon-pumped, phosphor-coated ultraviolet emitters for displays. The Japanese were expected to dominate flat-panel-display technologies "well into the future." Research support in II-VI (ZnSe) wide-band-gap lasers at U.S. universities was predicted to shift to support for nitride research. The Japanese were expected to dramatically improve the longevity and external efficiency of II-VI lasers and light-emitting diodes.

It is noteworthy that the terms "globalization," "outsourcing," and "internationalization" were not explicitly mentioned in the 2000 report's section on electronic and photonic materials. In 2004 the electronics industry and its materials supply chain were moving toward a global processing and manufacturing infrastructure. Large electronic materials suppliers had globalized their manufacturing base and support labs. Electronic original equipment manufacturers (OEMs) were globalizing R&D labs to support regionalized manufacturing operations. The rapidity of this change was not anticipated in the 2000 study, but there was an awareness that such changes were beginning.

In the 2000 report it was stated that the industrial strength of the U.S. materials science and engineering research community could not be compared meaningfully with that of a single country. The collaboration of university and industry researchers was noted as an important aspect of the U.S. innovation system. In particular, the benefit of individuals moving between the academic and industrial worlds was noted. It was further stated that the elimination of central research laboratories and of longer term, innovative research by many high-tech companies was making the technology transfer from universities more difficult. Today the U.S. electronics material industry's R&D is being globalized to support a worldwide industrial base. It remains too early to say if the weakening of the U.S. industrial R&D base will weaken U.S. innovation in the field of electronic materials or if there are alternative pathways for the transfer of innovative technology, such as industrial research consortia.

The 2000 report noted that the "valley of death" between innovation and application was becoming critical as development cycles became shorter. It further noted that proactively addressing this weakness was crucial to continued economic competitiveness in areas that depend on new materials. In the United States, industrial consortia such as Sematech and the International Electronics Manufacturing Initiative (iNEMI) have been formed to define research needs and to pull research results across that valley. However, as the electronic industry globalizes we often see American innovation being implemented into processes and manufacturing in other regions of the world, particularly Asia.

Today the United States is leading in R&D of materials and processes for semiconductor devices, particularly for evolving complementary metal oxide semiconductor (CMOS) technology. An example would be strained silicon technology. However, Japan and Korea are leading in the R&D of materials for displays and optical memories. In the area of organic printed wiring board materials, the R&D leadership is moving from the United States and Europe to Asia. A similar pattern is starting in materials for electronics packaging.

Semiconductor manufacturing is dominated by the United States, Europe, Japan, Taiwan, and Korea. Disk drive manufacturing is dominated by Singapore and Japan. Organic printed circuit board manufacturing for consumer products is moving to China from Japan, Taiwan, and Korea. Consumer electronics assembly has moved to China to be near the growing market and to take advantage of low-cost labor. In almost all cases, the global firms who provide materials are moving their development, manufacturing, and customer support functions close to the new manufacturing base.

Recently there has been rejuvenation in U.S. materials research to address the projected end of Moore's law and the scaling of CMOS devices. Many, including the International Technology Roadmap for Semiconductors (ITRS), anticipate that the traditional technology for cutting-edge devices will have fully evolved by 2015 and are looking for innovative solutions through new materials, new devices, or new architectures. The Semiconductor Research Corporation (SRC) has started a major new initiative in conjunction with the National Nanotechnology Initiative (NNI) and the NSF to focus on new nanostructures that might skirt this forecasted technology roadblock. These efforts will develop a new generation of researchers focused on new materials and structures for electronic devices. The iNEMI is road mapping the impact that these new nanodevices would have on the packaging and assembly of electronic devices. New electronic packaging will have to have micron-size leads and be capable of greatly enhanced heat transfer. Creative new material composites containing nanoparticles may provide the solution. They are being developed in many countries, and the technology is being bought or licensed from and by multinationals and start-ups around the world.

Another area of increased interest in the electronics community is sensors of all types: mechanical, fluidic, biological, chemical, radio frequency, and optical. Automotive, medical, consumer, and military end use electronics markets envision new products using inexpensive microsensors integrated with electronics processing. These sensors, called MEMs or "systems in package," will use a variety of silicon, ceramic, and organic substrates and call for new materials, particularly organic or biological. There is strong research leadership in the United States, followed by Europe, in materials for this growing area.

In summary, the evidence presented here and the literature data in Chapter 2 suggest the situation in electronic and optical-photonic materials is mixed, with the United States leading in some areas and not in others, often determined by commercial and market factors.

Superconducting Materials

- The 2000 benchmarking report concluded that the strong position of the United States in superconducting materials was not "assured." It noted that while U.S. industrial research in the field lagged behind Japan's, some small U.S. companies maintained world leadership in the design, manufacture, and characterization of long-length conductors, although the panel warned that "the shift in U.S. corporate research away from longer-term basic studies presents a question for the future." The panel noted that the momentum at that time favored relative improvements in the U.S. leadership position in some areas (magnetic properties, flux transport measurements and imaging, thin-film processing, and cable development), but without contin-

> ued strong federal investment in basic and applied research, that position would change. The panel concluded that the United States was poised, with strong processing and manufacturing capabilities and a growing talent pool, to capture a substantial segment of the superconducting market.

The mechanism for superconductivity in high-temperature superconductors (HTSCs) remains a very active research issue, although the technological breakthroughs made possible by the 1986 discovery are only beginning to be seen. There are both economic and military needs that are driving superconductivity research. In particular, the U.S. electricity supply system needs modernization and expansion to meet the needs of a growing economy and population. HTSC power technologies could play an important role, and electric utilities and large equipment manufacturers are planning R&D as well as installing and operating first-of-a-kind prototypes on their systems. Importantly, government funds are matched equally by private funds in the United States, which greatly increases the activity level. Similar pressures from growing electricity demand that might drive HTSC R&D here do not exist in Japan or Europe.

After it became apparent that wire technologies had to come first, the DOD reduced its R&D but closely tracked progress related to power applications. That changed with a 2002 determination by DOD that the second-generation (2G) wire being researched by DOE was "the critical component for several defense applications that require high electrical power and are essential to the national defense." Since then, DOD has participated with DOE in pilot-scale manufacturing for 2G wire. DOD also recently funded development of superconducting motors for ship propulsion and superconducting generators for airborne weapons use. The defense market, while relatively small compared with the commercial power market, is becoming an important driver of research.

The United States has been at the forefront in elucidating fundamental HTSC physical properties and has enjoyed leadership in synthesis and processing, with Europe and Japan also mounting strong efforts in these areas. Modeling of flux pinning and other phenomena that affect magnetic and electrical properties continues to be a vital area for research worldwide. The discovery of new HTSCs has slowed, and the materials discovered in the 1980s remain important worldwide, even though materials with higher transition temperatures (T_c) have been found. The grand scientific challenge has become to discover room-temperature superconductors, which would have enormous theoretical implications and would broaden technological applications beyond those now possible with HTSCs.

First-generation (1G) HTSC wires were successfully developed in the early 1990s and sold in the United States and Japan. European companies also developed manufacturing capability, and a Chinese company was recently established for this purpose. 1G wire is the only option now for power equipment development, but a 2G wire is expected to become available. The U.S. 2G leadership position is being challenged by a number of companies in Japan, Korea, China, and Europe. The United States continues to enjoy a leadership position in characterization and processing control technology for 2G wires and tapes. It has strong processing research capabilities at its national laboratories and excellent characterization facilities at DOE laboratories, universities, and the National Institute of Standards and Technology. Partnerships between government scientists and wire developers at U.S. companies increased recently, accelerating the availability of this important innovation. 2G wire research has been successful in many countries, and wires long enough to support power equipment development will likely be introduced in several countries before 2010. Applications needing higher magnetic fields—such as transformers, motors, and generators—are often put on hold to await the availability of 2G wires. Table 3.2 depicts the current status of 2G wire research worldwide.

DOE's Superconductivity Partnerships with Industry program has facilitated a number of first-of-a-kind utility-scale projects for several important electric power technologies. Japan and Europe have also had comprehensive electric power technology programs, and Korea and China recently began projects. The United States appears to have a commanding lead in planned installation of HTSC cables on the electricity grid.[12]

Fault current limiters are an important technology that is being pursued in several countries. The U.S. project, led by SuperPower, will test a 138-kV class fault current limiter in an American Electric Power substation. Japan was expected to discontinue all ac equipment projects at the end of 2004, including several on fault current limiters. The world's most powerful fault current limiter—a three-phase, 10-MVA device—was tested in Europe last year. There are motor and generator projects in the United States and Europe. The U.S. Navy is building HTSC shipboard propulsion motors with target ratings exceeding 30,000 hp. Rockwell Automation demonstrated a 1,500-hp motor and continues to develop the technology for larger class machines. Alstom is planning to operate a 250-kW machine in 2005. General Electric has built components for a 100-MVA-class generator but has delayed development pending availability of 2G wires. DOD is funding the development of high-speed HTSC generators for use on aircraft; these generators may use 2G wires. The U.S. transformer project led by Waukesha Electric Systems

[12]In 2005, HTSC cables are to be installed at two sites in New York and one site in Ohio, with operation scheduled in 2006 for all. Perhaps the most technically ambitious of the projects will be more than 650 m long and will operate at 138 kV, which is the most ambitious anywhere. Cable projects are either operational or planned in Japan, Korea, and China.

TABLE 3.2 Second-Generation HTSC Wire Technology, December 2004

Country or Region/ Organization	Length (m)	Critical Current at 77 K (A/cm width)	Substrate/ HTSC Deposition Process
United States/ American Superconductor	34	186	RABiTS/MOD
	10	272	
	Short samples	~400	
United States/SuperPower	100	70	IBAD/PLD
	62	100	IBAD/MOCVD
Japan/Sumitomo	35	175	RABiTS/PLD
	105	NA	RABiTS
	Short samples	357	RABiTS/PLD
	10	130	RABiTS/MOD
Japan/Showa Electric	230	NA	RABiTS (Ni-W)
Japan/Fujikura	105	126	IBAD/PLD
	Short samples	~300	
	255	NA	IBAD
Japan/ISTEC	45.8	182	IBAD/PLD
	8.6	119	IBAD/MOD
	220	NA	IBAD
	Short samples	413	IBAD/MOD
Europe/THEVA	1	422	ISD/Evaporation
	5	237	
	10	148	
Europe/Edison Spa	2	120	RABiTS/Coevaporation
	Short samples	>200	
Korea/KERI	4	97	RABiTS/Coevaporation
	1	107	RABiTS/PLD

was completed in 2004, with no follow-on device fabrication planned until 2G wire and appropriate cryogenic dielectrics became available. In Europe, Getra is building a small demonstration unit, and the Japanese have university-level activity.

In summary, U.S. scientists working on superconductivity are among the world leaders in nearly all component fields of superconducting materials. The United States, however, does not dominate in any, because other countries share or surpass the U.S. lead, often reflecting commercial realities.

Polymers

- The 2000 benchmarking report noted the United States had paid less attention and given less funding than many other countries to polymer research and that it "could lose ground in relative terms if not in absolute terms." The report noted the importance of polymer research to the U.S. positive trade balance in much of the chemical industry and that sustaining that balance would require the maintenance of U.S. world leadership in polymer research. The report also noted that environmental and life-cycle concerns were drivers for polymer research and development in Europe and were becoming drivers in the United States.

A review of polymer research in 2005 leads to an optimistic overall forecast for the field, consistent with the conclusions in 2000. There has been a resurgence of excitement over polymer science at a number of long-standing polymer departments at universities such as the University of Massachusetts at Amherst, Case Western Reserve University, and the University of Southern Mississippi, and many young faculty have been hired. In addition, researchers with a strong polymer science focus are now commonly found in the chemistry, chemical engineering, and materials science departments of leading U.S. universities such as the Massachusetts Institute of Technology (MIT), Cornell, the California Institute of Technology, the University of North Carolina, and the University of Texas. Similar trends are noted in Europe and Asia; the field is becoming more accepted by the classical science disciplines and is international in scope and distribution of skills.

The 2000 report also noted that much of the early research was done in U.S. industrial laboratories. This is becoming less and less the case: There is a strong trend at large U.S. polymer companies to substantially reduce the size of their early-stage research infrastructures and to refocus efforts to achieve shorter-term business/technical goals. A similar reshaping of the research infrastructure is also noted in Europe and Asia.

Long-term disruptive research (time to commercialization, 10 years or more) is being replaced by research with shorter-term goals (2- to 5-year programs or shorter). To an extent these trends are offset by research growth in the polymer user industry (computation, communication, medical devices) and the shifting of higher-risk, long-term research to start-up companies and university collaborations. While these trends suggest that significant levels of early-stage polymer research are continuing in the United States, the location of the engineering and scale-up infrastructure that can bring these ideas to commercial fruition, especially to serve large-volume markets, is unclear.

The 2000 benchmarking study noted the importance of the U.S. polymer industry for the positive balance of trade enjoyed by the U.S. chemical industry. That positive balance was seriously eroded in the first few years of the 21st century.[13] Table 3.3 summarizes the impact of the U.S. chemical industry on the balance of trade. From these data it is clear that while the chemical industry's balance of trade has been eroding, the plastics (polymer) sector is the one bright spot. Clearly, for this to continue, there needs to be a steady stream of new products identified and commercialized by U.S. producers.

The 2000 benchmarking report noted that interest in biological approaches, in reducing environmental impacts, and in biomedical applications of materials provided opportunities for the polymer industry. With the cost of oil increasing and the cost of sugar (from corn, cane, and the like) decreasing, there is an increasing interest in polymers derived from biomass rather than petroleum. Major efforts in this area have been announced by DuPont, and a joint venture was set up between Dow Chemical and Cargill to commercialize corn-based polylactides as commodity plastics. The impact of the recent announcement that Dow Chemical has sold its share to Cargill is unclear, but the reception for the product to date has been significantly stronger in Asia than in the United States or Europe. The lactide polymers were initially developed as erodible polymeric biomaterials by Johnson and Johnson in the 1970s, and the center of research into sugar-derived polymers continues to reside in the United States. Assuming current world pricing trends in oil continue, this area offers research promise in the 21st century. Bioderived and bioinspired polymers are also attractive emerging areas of polymer research where the United States enjoys a leadership position.

Overall, the United States maintains a leadership position in polymer research, although longer-term, applied polymer research is shifting from the industrial to the academic sector. The challenge will be not only to maintain research leadership but also to reestablish a U.S. polymer research infrastructure that can successfully commercialize emerging products.

Catalysts

- The 2000 benchmarking report concluded that the leading position of the United States relative to the rest of the world in the subfield of catalysts was "likely to lose ground as a result of the targeted funding aimed at growing capabilities in other countries." The panel warned that catalysis research could stagnate in the United States without "stronger, better equipped re-

[13]American Chemical Society, *Globalization and the Chemical Industry,* ACS Industry Pavilion (2002). Available at http://www.chemistry.org/portal/a/c/s/1/acsdisplay.html?DOC=industry\2002_global.html.

TABLE 3.3 Trends in the Chemical Trade (million $)

	Exports				Imports				Trade Balance			
Industry Subsector	2001	2002	2003	2004[a]	2001	2002	2003	2004[a]	2001	2002	2003	2004[a]
Organic chemicals	16,424	16,408	20,103	25,026	29,712	30,365	32,887	35,522	−13,288	−13,957	−12,784	−10,496
Inorganic chemicals	5,578	5,461	5,576	5,954	6,153	6,019	7,420	7,981	−575	−558	−1,844	−2,027
Plastics	18,485	19,380	21,069	24,791	10,401	10,760	12,161	14,084	8,084	8,620	8,908	10,707
Fertilizers	2,077	2,106	2,342	2,644	1,890	1,619	2,129	2,381	187	487	213	263
Pharmaceuticals	15,031	15,773	17,776	23,438	18,628	24,749	31,517	34,740	−3,597	−8,976	−13,741	−11,302
Cosmetics	5,825	5,871	6,558	7,396	3,750	4,235	5,611	6,805	2,075	1,636	947	591
Dyes and colorants	3,782	3,861	4,138	4,576	2,478	2,357	2,481	2,686	1,304	1,504	1,657	1,890
Other	12,382	12,347	12,987	14,514	5,927	6,167	6,858	8,027	6,455	6,180	6,129	6,487
Total	79,584	81,207	91,549	108,338	78,939	86,271	101,054	112,226	645	−5,064	−9,505	−3,888

NOTE: The data presented show that while the balance of trade in the chemical industry has been eroding, the situation for plastics is the exception, showing a growing excess of exports over imports.

[a]2004 estimate by *Chemical and Engineering News.*

SOURCE: Bureau of the Census.

> search centers where researchers can work together with common goals." The report predicted that the United States would remain a world leader for the production of chemicals through catalytic reactions in the most energy-efficient, safe, and environmentally compatible way. However, university-based research would continue to suffer relative to industry research unless better equipment became available. The panel noted the need for continued investment in catalysis research targeted at encouraging innovation and allowing U.S. industry to participate in the growth of emerging markets.

As a technology, catalysis has for decades provided American business with leadership positions, largely in the refining of petroleum to fuels, the production of a host of chemicals and polymers, power generation, emissions controls, and so on. Since the 2000 report there have been big changes in the catalysis industry—that is, catalyst producers and the chemicals and petrochemicals industries. Globalization has played a role in these changes in a number of ways. Global conglomerates that are no longer so heavily involved in the U.S. market are rushing to participate in the growth in Asia, especially China, now a major importer and exporter of chemicals as well as the number 2 importer of oil in the world. Catalyst producers have undergone substantial consolidation, as have many companies and industries that use catalysts.[14] The disruption in the U.S. industry as it becomes a commodity business has been accompanied by a reduction in the fundamental R&D these companies used to support or carry out. Catalysis research at companies in the United States has seen no significant growth since the 2000 report.

On the world scene, catalysis research continues to be a vibrant activity. Technical societies such as the Catalysis Society, with its regional organizations in the United States and abroad, continue to draw large numbers of scientists to meetings. Other forums, such as the American Chemical Society meetings and the American Institute of Chemical Engineering, are also active. Europe has developed outstanding centers of excellence. Highly respected catalyst R&D centers now exist in the Netherlands and in China.

[14]ICI's catalyst business, Synetix, most of Degussa's catalyst business, Activated Metals, Harshaw, Calsicat, and Houdry have become part of larger conglomerate catalyst companies. Chemical companies such as ICI, Hoechst, DuPont, and BASF have undergone major change. Chemical and petroleum companies such as Mobil (acquired by Exxon), Union Carbide, and Sohio no longer exist. DuPont sold its catalyst-intensive nylon business. No major new refineries have been built in the United States; most are being built in Asia.

A recent report by the U.S. International Trade Commission (ITC) concludes that catalysis in the United States has suffered greatly in the last 5 years.[15] Both American industry and universities have deemphasized catalysis research in the face of budget tightening and seemingly more glamorous technologies (biotechnology, electronics, and so on). As the ITC trade report comments, "a substantial percentage of expenditures by U.S. companies for contract catalysis R&D is now spent overseas, whereas staffing in company-funded catalysis has been reduced as a result of increased focus on the bottom line, which has also led to reduced U.S. university enrollment in catalysis-related R&D; although new technology development is accelerating, market application development has not always kept pace." The report also states that "U.S. universities are increasingly becoming a vehicle for research. Industry instead focuses on development and commercialization." The ITC concludes, "U.S. leadership role in catalysis has been eroding, but U.S. is highly competitive."

The foremost topics in catalysis continue to be environmental catalysis, fuel processing, selective oxidation, acid-base catalysis, biocatalysis, gas-to-liquids, production of hydrogen, asymmetric catalysis, photocatalysis, and chemical processing. Energy and the environment continue to be growth areas for catalysis R&D. Environmental legislation will continue to create a demand for new catalysts and catalytic processes. For example, regulations to reduce the sulfur content of gasoline and diesel fuel and limits on emissions from refineries will influence the refining catalyst business. Catalysis is seen by DOE to be critical to the development of a hydrogen economy. Catalysts are seen to provide the vital, crosscutting research needed, and they are viewed as "central to energy conversion."[16]

Research at the nanoscale promises to provide new understanding of the structure-property relationships for catalytic processes, while use of Operando (in situ characterization techniques)-driven research allows monitoring working catalysis under actual process conditions. These approaches, in combination with high-throughput screening, provide new techniques and methodologies for the discovery and development of new catalysts and more energy-efficient chemical processes.

In summary, catalysis R&D in the United States reflects the commercial reality. As many of the major industries that use catalysts have refocused attention on serving overseas demand, catalysis research in the United States has suffered. Since

---

[15]International Trade Commission, Report No. 3602, available at http://www.USITC.gov/ITTR.htm.

[16]DOE, *Basic Research Needs for the Hydrogen Economy* (2004), available at http://www.sc.doe.gov/BES.

the 2000 report, the United States has seen a continued decline in its former dominance in catalysis technology. The evidence presented here and the patent and literature data presented in Chapter 2 suggest the United States is steadily losing its leadership of this critical technology, going from the world leadership position in the 1970s and 1980s to one now of collective leadership. However, many exciting areas for catalysis research remain and could sustain a healthy research base in catalysis.

Nanomaterials

- The materials subfield of nanomaterials was not addressed as an individual subfield in the 2000 benchmarking report; in the section on electronic and optoelectronic materials, the report noted that in the field of nanotechnology, the United States had traditionally been a leader in exploratory nanostructures, including quantum wires and dots.

Of all of the areas of nanotechnology, the subarea of nanomaterials is widely expected to grow most rapidly. For one thing, the number of nanoparticle companies in the world increased from fewer than 20 to more than 200 between 2002 and 2004. Nanomaterial technology is moving faster than other nanotechnologies because of where nanomaterials fit in the product value chain. Figure 3.2, from the *Nanotechnology Opportunity Report*,[17] illustrates the differences between nanotechnology strategies from region to region. The United States has only a modest lead in nanomaterials, a lead that is not likely to last long. This situation can be attributed to the nature of nanotechnology innovation in the United States compared with other parts of the world. In fact, the United States is the near-term innovation leader, with the largest investment in start-ups and small companies, and this position gives it an early lead in generating intellectual property. However, although entrepreneurial activity in the United States is impressively fast-paced, it might turn out to be inefficient, because the nanomaterials technologies have not been developed in ways that are suited to small companies or start-ups. Meanwhile, in Europe, greater investment is being directed toward fundamental R&D at universities and other research institutions, a strategy that may bring more significant breakthroughs in nanomaterials design synthesis in the long term. Asia—China, Japan, and Korea in particular—has adopted and integrated nano-

[17] See Científica, *The Nanotechnology Opportunity Report* (2003), available at http://www.cientifica.com/html/NOR/NORV2.htm. The report profiles over 800 nanotechnology companies, research groups, and investors worldwide.

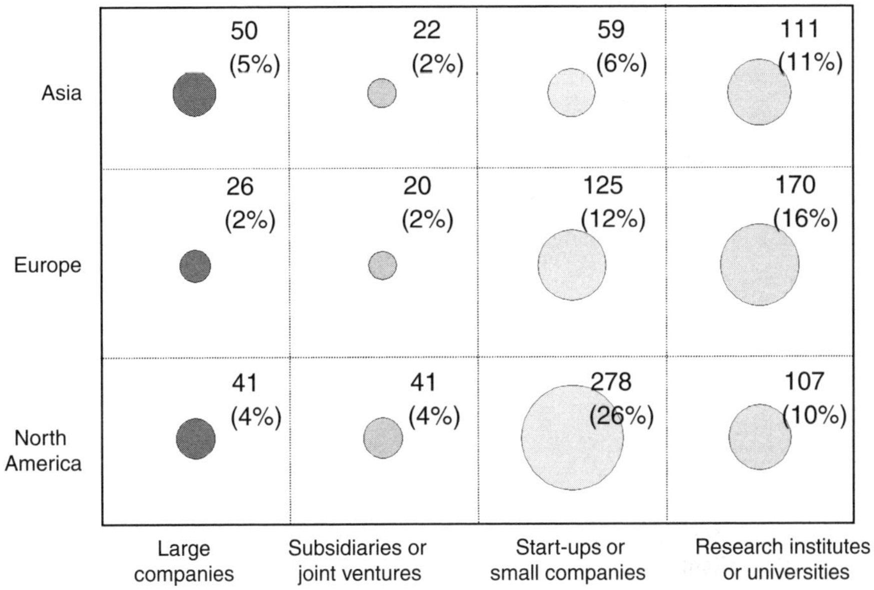

FIGURE 3.2 Global competition. Players by type in competing economic blocs in the nanotechnology sector. Size of the circle is proportional to the number of actors in the field. SOURCE: Científica and Jaakko Pöyry Consulting.

technology research into large commercial enterprises. The Asian strategy may ultimately lead to profitable manufacturing of nanostructured materials.

Figure 3.3, also from the *Nanotechnology Opportunity Report*, shows the distribution of nanotechnology R&D among various kinds of organizations. Table 3.4 shows the growth in select nanomaterials research areas, according to the cumulative number of publications found using the Los Alamos National Laboratory FlashPoint Multidatabase literature search tool. Literature analyses such as these reveal the maturity of nanomaterials research and point to some of the hot topics. Growth has been relatively uniform across the areas of nanomaterials research, although some more mature areas are growing more slowly.

Clearly there is ongoing and growing excitement about nanoscience in the international research community, but the sources of financial support for nanotechnology indicate how the resources are being channeled into the development of commercial products. It is largely through commercialization that nanomaterials and nanotechnology will impact quality of life and economic prosperity.

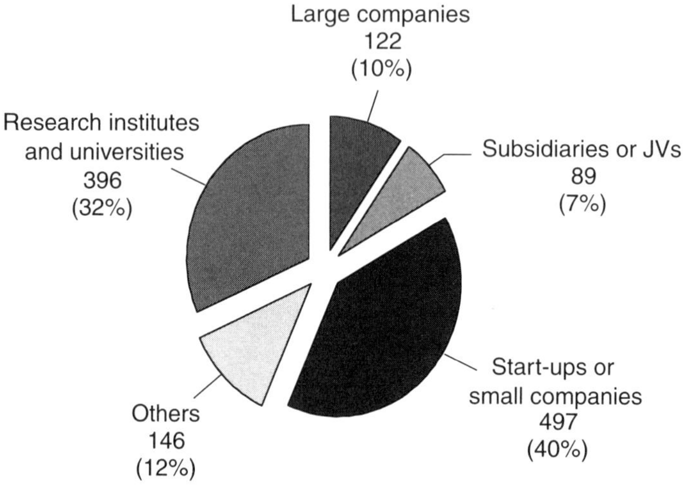

FIGURE 3.3 Distribution of entities contributing to nanotechnology development in the world. SOURCE: Científica and Jaakko Pöyry Consulting.

All sources of capital for nanotechnology are growing, with the global pattern for this growth one indication of where nanotechnology is headed. The United States remains the global leader in the number of corporations engaged in nanotechnology, as shown in Figure 3.4. The corporate investment in nanotechnology worldwide totaled $2.0 billion in 2001; by 2003, it had exceeded $2.8 billion. In the same time period, government investment increased from about $2.0 billion to $3.0 billion. Government investment in Japan in nanotechnology research is growing more rapidly and catching up with analogous U.S. and European investments. Figure 3.5 shows that venture capital is the most significant source of support for

TABLE 3.4 Growth in Select Nanomaterials Research Areas

	Literature Search Term		
Year	Nanoparticle	Nanotube	Quantum Dot
2000	712	837	3,346
2001	1,075	1,449	3,755
2002	1,692	2,057	4,259
2003	2,372	2,920	5,010

NOTE: Based on the cumulative number of publications found using the Los Alamos National Laboratory FlashPoint Multidatabase literature search tool.

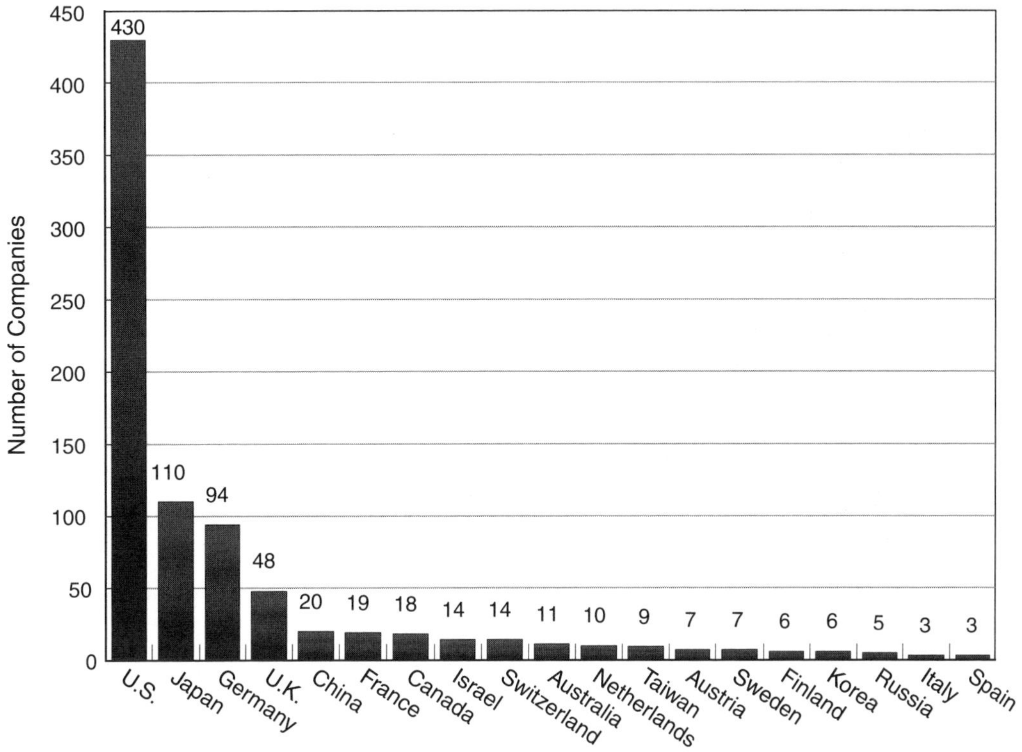

FIGURE 3.4 Companies active in the nanosector, grouped by country of origin. SOURCE: Científica and Jaakko Pöyry Consulting.

nanotechnology in Western countries such as the United States, the United Kingdom, and Germany. The venture community is playing a critical role in the growth of nanomaterials commercialization in the United States. In contrast, corporate investment dominates nanotechnology commercialization in Japan. We can therefore expect the incorporation of nanotechnology in existing corporate products in Asia, while in the United States we should see a greater proliferation of novel nanomaterials and nanotechnologies that come from highly innovative start-up companies.

In summary, while the United States leads global activity in nanomaterials and nanotechnology as measured by the number of corporations engaged in the subfield, it is too early to say which region of the world, if any, is going to show clear leadership as this field matures. The use of nanotechnology in many electronic and optical-photonic materials and devices means the U.S. position in both MSE subfields will remain interconnected.

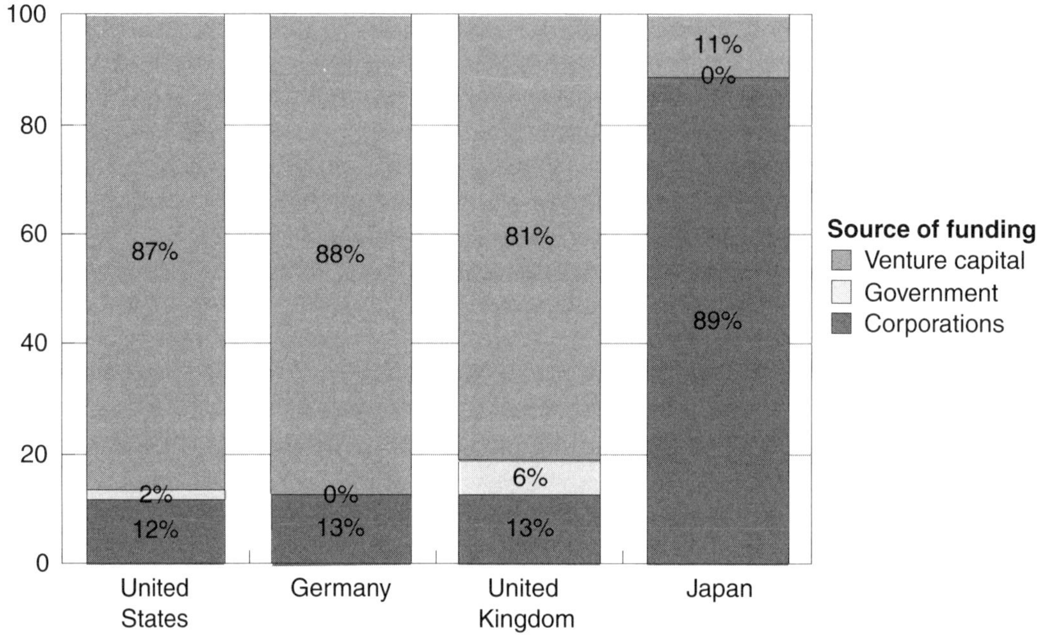

FIGURE 3.5 Sources of capital for nanotechnology in four countries, through end 2002. SOURCE: Científica and Jaakko Pöyry Consulting.

CURRENT STATUS OF MSE R&D

According to the benchmarking reported here, the United States is still the leader or among the leaders in research in fields where the domestic process or manufacturing industry is leading or strong. It appears that when an industry begins to move to other regions of the world, there are two different models: (1) the growth of an industry outside the United States leads to a decline in domestic research and (2) a proactive industry forms a strategy, and the strong materials R&D base in the United States is sustained. Catalysis is an example of a subfield where U.S. research declined along with the global growth of the petroleum and chemical industry and its relative decline in the United States. Electronic materials for CMOS and magnetic disk drives are examples where the manufacturing industry moves to another region but the industry has acted proactively to maintain the strong materials R&D base in the United States.

The benchmarking evidence in this chapter and the patent and literature evidence from Chapter 2 make it clear that the increasing international and transnational activity in MSE R&D is impacting U.S. MSE R&D and its standing in the world. In this regard, the committee offers two conclusions.

Conclusion. Globalization of MSE R&D is proceeding rapidly, in line with broader trends toward globalization. As a result of increasing international trade and investment, the emergence of new markets, and the growth of the Internet and the global communications system, MSE R&D in the United States is an internationalized activity with a diverse set of international partners.

Conclusion. The globalization of MSE R&D is narrowing the technological lead of the United States.

4

The Regulatory Regime as a Driver

Regulatory regimes serve a number of purposes: They can govern business and trade, control information of national importance, ensure that inventors own the results of innovation, determine ethical practices, ensure the safety of employees, regulate the migration of workers across international borders, and provide a framework for the execution of international relations.

This report discusses regulatory regimes that may affect the globalization of R&D activities. While international treaties on scientific cooperation can clearly drive new R&D activity—such as increasing the participation of U.S. researchers in European Union (EU)-funded projects, as happened following the completion of the EU-U.S. 1997 Agreement for Scientific and Technological Cooperation—the influence of regimes that regulate exports, taxes, intellectual property, and so on is less clear. This chapter considers some of the regimes likely to influence the corporate decision-making processes that are associated with globalizing R&D. The analysis of the regimes herein is not meant to advocate for or militate against particular regulations; rather, it is intended to show how the regulations might influence globalization decisions.

EXPORT REGULATION AND TECHNOLOGY TRANSFER

Bayh-Dole Act

The Bayh-Dole Act, enacted in 1980 as an amendment to the Patent and Trademark Act, applies to the transfer of university-generated, federally funded inventions to the commercial market. The legislation was designed to address problems with the ownership of technology and inventions resulting from federally funded research performed at colleges and universities and the apparent reluctance on the part of industry to invest in the further development and manufacture of technology that rarely made it to market. The Bayh-Dole Act allows universities and small businesses to take title to the inventions, encouraging further development and manufacture. There are two requirements of the Act. First, the invention must be patented in order for the university or small business to take title to it. Second, any manufacture of the product must be substantially performed in the United States. A waiver to this requirement can be obtained if, after a reasonable search, no manufacturer can be found or if manufacture in the United States is economically unfeasible.

These requirements can hinder meaningful international participation, and once the result of the research is patented, the related research is no longer basic. If the technology is subject to export controls, this affects further R&D in the United States, because any students involved would have to be either U.S. citizens or their work covered by an export license or other export authority. More significantly, the Act prohibits manufacture outside the United States, which makes corporations unable to consider offshore outsourcing for technology covered by the legislation. Critics of this aspect of the legislation say it effectively puts potential developers or users of federally funded university research in the same situation they were in before 1980.

The end result of these requirements is that industry might be reluctant to engage in federally funded industry-academia alliances in the United States and more favorably inclined to consider such alliances with non-U.S. universities funded by foreign governments, subject to local regulations of course.

Export Regulations

The primary sources of export regulation—the Department of State's International Traffic in Arms Regulations (ITAR) and the Department of Commerce's Export Administration Regulations (EAR)—are considered by some in industry as a barrier to the global conduct of business. To compete in the global market and maintain a comparative advantage, U.S. industry must have access to both domestic and foreign technology, and manufacturing and export controls could be considered as hindering this access. Critics of the current export regulation regime maintain that foreign companies are executing contracts while U.S. companies are still seeking regulatory approval.

Over the past 20 years most congressional activity on the export regimes has been to add sanctions and restrictions rather than to substantively review the underlying statutes. Following the terrorist attacks of September 11, the ensuing focus on national security, and the further regulation of the transfer of items and technology that could be related to weapons of mass destruction, the emphasis has remained on control of technology rather than on easing its transfer. The two primary sources of regulation provide further insight.

International Traffic in Arms Regulations

ITAR applies to items on the Munitions Control List—that is, to military end items, components, and the underlying technical data. In all cases a license or other authority is required prior to any export. The origins of the ITAR can be found in the Neutrality Acts of the mid-1930s. The underlying policy, then and now, is control over the transfer of arms and related technology. The Cold War was a major factor in shaping regulations as well as their interpretation. Ongoing multiagency review of the items on the list has resulted in some items being transferred from Department of State jurisdiction to the Department of Commerce.

Because the focus is on control, reviews of applications for export license or other export authority have no set time limit. Applications may take four or more months to complete. Approval is by no means assured and may be accompanied by conditions and limitations. The result is that where export approval is required, U.S. industry can find itself unable to plan with certainty, because there is no way of knowing when approval may be granted or how the license provisos may impact planned performance.

All of the above must be considered before entering into ITAR-controlled research projects with foreign companies. While ITAR is clearly critical for protecting the nation's interests in the systems and knowledge it covers, the ITAR regime can lead to schedule uncertainties, cumbersome regulatory requirements, and compliance risks that inhibit international collaboration with U.S. suppliers and partners.

Export Administration Regulations

EAR applies to commercial and dual-use commodities and their materials, components, software, and technology. Applying EAR is a multistep process. First, it must be determined whether the commodity is controlled under the EAR, no matter whether the commodity or its underlying technology is being exported. Next, the control regimes applicable to the export—national security, nuclear nonproliferation, and so on—must be determined. Finally, it must be determined whether the control regime applies to the country or countries for which the export is intended.

The underlying statutory authority, the Export Administration Act, dates back to 1979. While that legislation lapsed in 1994 and annual attempts to update and reinstitute the Act have failed, the regime has been kept alive by Executive Order and annual notices. It lapsed again in 2001 but was revived under the International Emergency Economic Powers Act. The 2003 attempt to pass legislation failed in large part because it was deemed to not sufficiently strengthen national security controls on exports. As a result, the Department of Commerce is working to increase administrative controls on knowledge/technology transfer and exports. Critics say that the EAR impedes the collaborative efforts necessary for the conduct of global R&D.

Export Regulation and Technology Transfer (Offsets)

Industrial participation, or "offset," is another path for the globalization of R&D. Offset obligations are incurred when an agency of a foreign government purchasing a military or commercial product requires that some reciprocal activity take place to help offset the import/export imbalance created in its country by the purchase. One way to accomplish this is to transfer technology or intellectual property to the purchasing country's industries. Critics of offsets point out that they can have an adverse impact on domestic U.S. suppliers, which are perceived as being forced to provide foreign companies with their technology, resulting in a loss of domestic business, jobs, and know-how.

Any transfer of technology or intellectual property typically occurs in one of two ways. The first is a one-way transfer by which the recipient organization is provided training, data, software, or some other intellectual property that enhances its knowledge and capabilities in a specific technological area. The second way is technology collaboration, a two-way transfer of technology in which the companies typically share intellectual property to develop a specific product or technology. In either case, since technology is being transferred out of the United States and into a foreign country, that technology or intellectual property may be subject to ITAR or EAR.

If the technical data are inherently military or can be applied to a military platform in the recipient country, they will be subject to ITAR and their transfer (or their use in a collaboration) will require a review and license from the Department of State. If the technical data are dual-use, military and commercial, then they will typically be subject to EAR and will require a review and license from the Department of Commerce. The type of license required will depend on whether the technical data are a permanent or temporary export/import and whether the service provided constitutes technical or manufacturing assistance. Either set of regulations (ITAR or EAR) can impact the extent to which a transfer or collaboration across borders takes place.

The greatest impact would be if the transfer or collaboration is denied. A denial might be the result of U.S. government disclosure policy relative to a specific country in which the offset is proposed or of the category in which the technical activity falls, or of a combination thereof. Even if a license is awarded, provisos or limitations are usually placed on the offset activity that can greatly constrain the technology transfer or collaboration. These licenses and provisos can impede global research activities by inhibiting the necessary sharing of intellectual property and results of the research with those in the collaboration. Typically, the process to secure a license through the Department of State or the Department of Commerce is lengthy. This may delay progress and increase costs, squandering the leverage gained by collaboration. In addition, since most of these licenses are very specific about the technical activity that can be performed, any deviation from the description in the license would require another license to be obtained. Critics say that because the path often is not clear when the first step of the R&D is taken, these regulations limit the creativity and robustness generally expected in a vibrant R&D program.

Tariffs

Free trade agreements, whose proponents extol the benefits of lowering or limiting tariffs to trade, have proliferated over the past few years. It is estimated that several hundred such agreements now exist. U.S. law does not typically impose tariffs on imports of technology. However, to the extent that a U.S. company has assisted a foreign supplier by providing technology or data or services that are integrated into the item to be imported, that "assist" is valued in determining a tariff. In general, however, tariffs are not considered a barrier to the global conduct of R&D as they do not come into play until actual commodities are moved across borders.

INTELLECTUAL PROPERTY LAW

Intellectual property (IP) rights are government-granted and -protected rights in innovation and creativity. Because IP rights are creatures of geography, that is, they are national in scope, the protection and enforcement afforded can vary considerably from country to country. Performing R&D outside the United States therefore risks that IP created or used in that activity might not be respected, and that the rights in it could not be enforced effectively to the extent that they are under U.S. law. Early international agreements did not resolve these concerns. For example, the Paris Convention of 1883 protects against trademark and patent infringement but does not mandate any substantive intellectual property standards in local law. The Berne Convention of 1886, which provides protection against copyright infringement, does not provide any legal remedy for infringements.

In recent years, the situation has begun to change. The World Intellectual Property Organization (WIPO) was formed in 1974 as an agency of the United Nations with a charter to promote the protection of intellectual property throughout the world and facilitate the transfer of technology from developed to developing countries. The Trade Related Aspects of Intellectual Property Rights accord (TRIPS), signed on April 5, 1994, provides for a number of protections and obligations relating to intellectual property. Among other things, it (1) incorporates the substance of the Paris and Berne conventions and (2) requires (a) that member countries of the World Trade Organization have certain minimum protections in their local laws and (b) that those protections apply equally to nationals of other member countries.

Despite these changes, differences remain, and even in countries with well-established intellectual property regimes, differences in laws can affect the decision whether to do R&D in the United States or abroad. As an example, the laws in various European countries impose compulsory licensing obligations on patent holders who do not "work" an invention—that is, manufacture a patented product or practice a patented process within a certain period after patenting. Similarly, compulsory licenses can be required in certain countries in situations involving dominant, or blocking, patents and where such licenses are deemed to be consistent with the national interest.

U.S. law can limit the ability of domestic companies to rely on materials research conducted abroad. One of the main categories of limitations arises under U.S. patent law, Chapter 35 U.S.C. The patent laws generally provide a patentee with the right to exclude others from making, using, offering to sell, or selling any patented invention within the United States or importing into the United States any patented invention.[1] There is no exemption under U.S. law for the use of patented materials or processes in R&D connected to commercial activity.[2] The fact that research was carried on in a country where the work did not infringe another party's IP rights[3] does not provide an exception to the U.S. patent laws.

[1]35 U.S.C. § 271(a).

[2]Cases interpreting the patent statutes have established a limited experimental use exception for such use not connected with commercial activity.

[3]This would be so, for example, where no patents covering the activity had been issued in that country.

Therefore, the results of R&D activities conducted abroad in a country where no patent covered such work might still infringe a U.S. patent owner's rights if the results are utilized in the United States. However, infringement is limited to physical goods manufactured by a patented process and does not extend to information derived from such a patented process. In addition, under U.S. patent law, importation and use of a material made abroad by a process covered by a U.S. patent can itself constitute an infringement of the process patent, regardless of whether the alloy or material itself was a patented composition.[4]

TAX POLICY

The so-called Section 41 R&D tax credit is intended to serve as an incentive for the conduct of certain types of product development research activities and certain basic research. The credit is an incremental credit equal to the sum of 20 percent of the excess of qualified research expenditures for the taxable year over a base amount and 20 percent of basic research payments. Qualified research expenses include in-house expenses for wages paid and supplies used in the conduct of qualified research and 65 percent (75 percent in certain instances) of any contract expenses for qualified research. An alternative three-tiered research tax credit is available to generate higher research credits for companies with dramatically increasing sales figures or otherwise stagnant research expenditures.

In 1988, the U.S. Congress supported the Section 41 R&D tax credit when it stated: "Research is the life-blood of our economic progress and effective tax incentives for research and development must be a fundamental element of America's competitive strategy."[5] Supporters of the tax credit justify it because of the economic growth and productivity gains that come from a combination of technical progress and the associated investments in tangible capital assets, R&D, human capital, and public infrastructure. Under Section 41, only those costs associated with research performed within the United States are eligible for the credit, and the majority of the expenses claimed are for wages paid to U.S. employees. On October 4, 2004, President Bush signed into law the Working Families Tax Relief Act of 2004, which extended the Section 41 R&D tax credit retroactively, from after June 30, 2004, through December 31, 2005.

---

[4]Whoever without authority imports into the United States or offers to sell, sells, or uses within the United States a product which is made by a process patented in the United States shall be liable as an infringer, if the importation, offer to sell, sale, or use of the product occurs during the term of such process patent (35 U.S.C. § 271(g)).

[5]H.R. Rep. No. 100-1104 (1988), Conference Report on the Technical and Miscellaneous Revenue Act of 1988, at 88.

OTHER REGULATORY REGIMES

The regulatory regimes above are thought to be the regimes most likely to influence corporate decisions related to the globalization of R&D. There are, however, other regulations that impact the execution of R&D. These are summarized below.

Immigration

Since the terrorist attacks of September 11, 2001, the requirements of entry procedures for foreign visitors and their enforcement have been tightened, with unintended consequences for American science, engineering, and medicine. According to evidence collected from the U.S. scientific community on behalf of the Presidents of the National Academies,[6] ongoing research collaborations have been hampered; outstanding young scientists, engineers, and health researchers have been prevented from or delayed in entering this country; and important international conferences have been canceled or negatively impacted.

The United States has benefited enormously from the influx over the years of foreign-born scientists and engineers whose talents and energy have driven many of its advances in scientific research and technological development. In addition, the influx of foreign talent has increased the diversity of the U.S. research community, giving it an advantage in terms of identifying and applying new technologies from around the world. Foreign students are essential for much of the federally funded research carried out at academic laboratories. International conferences, collaborative research projects, and the shared use of large experimental facilities are essential for progress at the frontiers of these areas. The ongoing annual shortage of H1-B specialty worker visas is a concern for the high-technology sector. Further, as already seen, the MSE field is heavily reliant on foreign graduate students.

Environmental and Safety and Health Standards

Assuring environmental and health and safety compliance and the associated risk-reduction efforts is a challenge for U.S. industry.[7] Statutes, regulations, and

[6]See statement *Current Visa Restrictions Interfere with U.S. Science and Engineering Contributions to Important National Needs*, Bruce Alberts, President, National Academy of Sciences, Wm. A. Wulf, President, National Academy of Engineering, and Harvey Fineberg, President, Institute of Medicine, December 13, 2002 (revised June 13, 2003).

[7]Appendix G provides a brief but detailed overview of environmental, and health and safety regimes described in lesser detail here.

requirements established by international, federal, state, regional, and local agencies seek increasingly detailed reporting. Regulatory enforcement is also on the rise. Poor environmental and health and safety performance can have a significant impact on the bottom line as well as on customer and public perception.

It is important to understand which materials, processes, and equipment are or will be subject to environmental- or health/safety-driven limitations. This is especially critical for proprietary and R&D programs. Environmental regulations may sometimes exempt laboratory operations and R&D activities, but these exemptions typically do not apply when the activity is transitioned to production. In addition, waste, wastewater, air emissions, worker exposure, and so on from these activities must still meet local, state, and federal regulatory requirements and customer requirements. International restrictions are becoming more important as the global market expands and product customers are found throughout the world. The United States is also bound by international treaties such as the Montreal Protocol,[8] which may impose additional and, at times, more stringent restrictions than those already in force here. Relevant regimes on the environmental side include the Clean Air Act (CAA), the Clean Water Act (CWA), the Resource Conservation and Recovery Act (RCRA), and the Comprehensive Environmental Response, Compensation and Liability Act (CERCLA). Other laws to be aware of include the Emergency Planning and Community Right-to-Know Act (EPCRA), the Superfund Amendments and Reauthorization Act (SARA), the Toxic Substances Control Act (TSCA), and the National Environmental Policy Act (NEPA). On the health and safety side, the Occupational Safety and Health Administration (OSHA) promulgates standards, rules, and regulations for specific activities as well as hazardous and other material usage. For more information on these regulatory regimes, see Appendix G.

Product Approval Regulation

Materials approval regulation is defined herein as the collective body of documented criteria that dictate acceptability of materials from the standpoint of the product user. Such documentation typically takes the form of a published specification that serves as a common standard. These specifications can establish standards for "form" (composition, shape, and processing used during manufacture) and "function" (performance during intended use). Specifications are generated

[8]The Montreal Protocol on Substances That Deplete the Ozone Layer is an international agreement designed to protect the stratospheric ozone layer by controlling the production and consumption of compounds such as chlorofluorocarbons, halons, carbon tetrachloride, and methyl chloroform.

by corporations, industries, governmental agencies, and international organizations. The materials specifications most relevant to global discussions are those controlled by international organizations, such as the Society for the Advancement of Material and Process Engineering, the Society of Automotive Engineers, the International Organization for Standardization, ASM International, and ASTM International.

R&D of materials for use by the U.S. military must consider the acceptance criteria of the end user—one or more branches of the Air Force, Navy, Marines, and Army. These entities have authority over the materials requirements of the weapons systems manufactured by industry. Before either original equipment manufacturers (OEMs) or end users can implement a novel material on a defense product, the materials must be proven to conform to military specifications, commonly denoted by the prefix MIL. OEMs sometimes establish their own material specifications to suit the unique needs of a weapon manufacturer. Both OEM and MIL specifications routinely reference specifications owned by international organizations. Sometimes conformance to an OEM or MIL specification is not required, but industry convention or company demand for standardization will drive the use of materials organization specifications. Materials that successfully meet all stated requirements are generally added to a qualified products list (QPL). Items on the QPL for a given material specification are then deemed acceptable by the end user.

The trend with the most far-reaching impact in the context of this report is the increasing reliance upon standards with a more global reach. As multiple tiers of international suppliers work in concert to deliver specialized components for highly sophisticated finished products, common materials specifications are necessary to communicate standards throughout a complex, networked supply chain. Because it would result in fewer total procedures, the standardization of specifications could foster greater efficiency.

SUMMARY

In summary, this chapter has provided a snapshot of some of the regulatory regimes most relevant to the decision-making process associated with the globalization of R&D. The importance of these regimes as a driver of decisions by industry, government, and academia on locating R&D is likely to vary from decision to decision. It seems clear, however, that questions about the security of IP developed abroad, whether the export license process will hinder the execution of an R&D program, whether workers will be available to carry out the research, and what tax incentives might be in place in the locales under consideration would have to be answered in any thorough decision-making process. It is unlikely, however, that any single issue would be the sole determinant of the location of an R&D activity. Nevertheless, it is incumbent on the makers of public policy to make sure that as R&D becomes more global, U.S. researchers are not unreasonably impeded from taking part in international activities of national importance.

5

Assessing the Impacts of Materials Science and Engineering R&D Globalization

MSE R&D becomes globalized through a variety of mechanisms: domestic R&D activities, academic exchanges, international trade,[1] and foreign direct investment, as well as associated spillovers.[2] All the evidence indicates globalization of MSE R&D is well under way, so the question is how the United States should react to this phenomenon. Responding to this question requires an understanding of the possible impacts of the phenomenon.[3]

ECONOMIC IMPACT

Background

Assessing the economic impact of MSE R&D globalization is a challenging task given the complexity of the economic forces driving globalization and the inherent uncertainty surrounding economic outcomes. Quantifying the impact of

[1]R&D is considered a business service in national and balance of payments accounts.

[2]The concept of spillovers is that one invention might speed up other inventions or a country might acquire technological knowledge through imports of intermediate and capital goods. Spillovers can result in a situation where R&D investment creates benefits not only for the inventor but also for others.

[3]The charge for this study limits its scope to the impacts of the globalization of R&D in the MSE sector. It does not call for analyzing the impact of the globalization of the MSE sector as a whole—that is, from R&D, to processing, to manufacturing.

BOX 5.1 Economic Impact of R&D Outsourcing

A few studies estimating the economic impact of outsourcing appear to be directly relevant to the study of globalization of MSE R&D. Most of them were carried out using European data and focused more on the outsourcing of intermediate goods and not so much on services.[a] The most well-known analysis for the United States[b] focused on the outsourcing of intermediate inputs and its effects on the wage premium for skilled workers. The studies did not, however, consider outsourcing of services. A back-of-the-envelope estimate for the IT industry[c] suggests that the potential benefits of outsourcing services are high and concludes that IT outsourcing led to an annual increase in productivity of 0.3 percent points from 1995 to 2002, or $230 billion in additional GDP.

European studies[d] used plant-level data for the electronics industry in Ireland from 1990 to 1995 to show that international outsourcing of services had a positive, albeit not very

[a]Several studies identify cost-saving as the primary motive for outsourcing intermediate goods, among them, Hartmut Egger and Peter Egger, "International Outsourcing and the Productivity of Low-Skilled Labor in the EU," *WIFO-Working Paper* 152 (2001), available at http://publikationen.wifo.ac.at/pls/wifosite/wifosite.wifo_search.frameset?p_filename=WIFOWORKINGPAPERS/PRIVATE5397/WP152.PDF; Hartmut Egger and Peter Egger, "Outsourcing and Skill-Specific Employment in a Small Economy: Austria After the Fall of the Iron Curtain," *Oxford Economic Papers* 55(4) (2003): 625–643; and Sourafel Girma and Holger Görg, "Evaluating the Causal Effects of Foreign Acquisition on Domestic Skilled and Unskilled Wages," Bonn: IZA-Institute for the Study of Labor (2003).

[b]Robert C. Feenstra and Gordon H. Hanson, "Global Production Sharing and Rising Inequality: A Survey of Trade and Wages," National Bureau of Economic Research Working Paper No. 8372 (2001). Feenstra and Hanson construct industry-by-industry estimates of outsourcing (of intermediate products) from 1972 to 1992. Looking at these data, they find that outsourcing contributed substantially to an increase in domestic demand for high-skilled nonproduction workers and in their wages.

[c]Catherine L. Mann, "Globalization of IT Services and White Collar Jobs: The Next Wave of Productivity Growth," *International Economics Policy Briefs* 3-11: Institute of International Economics (2003), available at http://www.iie.com/publications/pb/pb03-11.pdf.

[d]Holger Görg, Aoife Hanley, and Eric Strobl, "International Outsourcing and Productivity: Evidence from Plant Level Data," *GEP Research Paper 04/08,* Nottingham: University of Nottingham (2004), available at http://www.nottingham.ac.uk/economics/staff/details/papers/holgerweb7.pdf.

MSE R&D globalization requires data gathered over time on (1) domestic and international R&D in various MSE subfields and (2) the contribution of MSE R&D to the production of other goods and services in the economy. Such data are not readily available at present. The objective of this study is therefore more modest: to identify the key factors that are likely to determine the impact of materials R&D globalization on the U.S. economy. Understanding such factors is crucial to making an informed decision about whether any policy intervention is needed and, if so, what form it should take.

robust, effect on productivity growth.[e] Likewise, Girma and Görg[f] find a positive impact for outsourcing of industrial services[g] on productivity in the U.K. manufacturing industries from 1980 to 1992, although they are unable to distinguish between international and domestic outsourcing.

Turning more narrowly to the issue of R&D, it appears that R&D outsourcing abroad is still relatively small for the subfields considered. However, based on the benchmarking analysis and supported by the analysis of patent data and case studies, the United States seems to be losing the technological lead in some materials subfields. Europe and Japan are already strong competitors for the United States in a number of subfields, and emerging-market countries such as China and India, while not yet major players across the spectrum of MSE R&D, have the potential to become so in some subfields.

[e]The authors find that international outsourcing generally had a positive effect on productivity, most of which could be attributed to outsourcing of material inputs. Similarly, P. Egger, M. Pfaffermayr, and Y. Wolfmayr-Schnitzer, "The International Fragmentation of Austrian Manufacturing: The Effects of Outsourcing on Productivity and Wages," *North American Journal of Economics and Finance* 12 (2001): 257–272, find outsourcing of material inputs by Austrian manufacturing firms to Eastern transition economies increases domestic growth in total factor productivity, more so in capital-intensive industries than in labor-intensive ones. Egger and Egger (2003) find that a 1 percent increase in outsourcing of intermediate inputs to East European countries relative to gross production induces a shift in relative employment by about 0.1 percent in favor of high-skilled labor. Egger and Egger (2001) find that outsourcing of intermediate products by EU manufacturing firms reduces the productivity of low-skilled workers in the short run and increases it in the long run, an effect the authors attribute to imperfections in the EU labor and goods markets. (An updated version of Egger and Egger (2001) is in press at the time of finalization of this report for publication in *Economic Inquiry* in 2005.)

[f]S. Girma and H. Görg, "Outsourcing, Foreign Ownership, and Productivity: Evidence from UK Establishment-Level Data," *Review of International Economics* 12(5) (2004): 817–832.

[g]This study defines "industrial services" as "activities such as processing of inputs which are then sent back to the establishment for final assembly or sales, maintenance of production machinery, engineering or drafting services, etc." They do not include nonindustrial services such as accounting, consulting, cleaning, or transportation. It is not clear where R&D would fit under this definition. The Girma and Görg paper (2003) provides some estimates on the extent of outsourcing by industry.

Some Key Economic Factors

Over the last 20 years or so economists have carried out numerous studies that are helpful when considering the impacts of the globalization of MSE R&D (see Box 5.1, "Economic Impact of R&D Outsourcing)". The analysis frameworks developed in the course of these studies focus on R&D-based economic growth, international technology diffusion, economic catch-up, international trade, and foreign direct investment.[4] The conclusions of these analyses point to some key factors that are likely to determine the impact of MSE R&D globalization on the U.S. economy.

Innovation is an engine of economic growth. Along with labor and capital, it is a key source of growth in the long run. A country with a large stock of human capital and heavy investment in R&D activities has the potential to develop a comparative advantage in high-technology products, to capture a large share of the markets for them, and to run a trade surplus in this category of products. Not only the innovating country but also its trading partners can benefit from the increasing variety and higher quality of products available through international trade.

The global knowledge economy is fueled by an ever-faster information flow, accelerating knowledge generation, and the emergence of new, highly networked ways to handle and disseminate information and knowledge. Almost all industries and sectors have been and are being affected, broadly and specifically, by the globalization in technology. In broad terms, the global landscape can be characterized by shorter R&D and product life cycles, greater productivity, and the consolidation and globalization of production. Globalization of R&D is just one facet of a broader globalization phenomenon exemplified in increasing international flows of goods, services, capital, people, and information.

Government policies on trade, on R&D, and on production subsidies can affect international patterns of specialization, trade, and technology diffusion. In some cases, R&D subsidies—and, more generally, incentives associated with national innovation strategies—might help promote new R&D activities in a country that otherwise would not have had them and create a comparative advantage in a particular R&D area. In other cases, however, R&D subsidies might be counterproductive, because they shift scarce human resources away from more productive activities, such as the manufacture of technology-intensive products.

International technology diffusion—whether through trade, direct foreign investment, or more indirect means—can enhance innovation and economic growth by fostering competition in R&D and by providing successful innovators with greater access to markets. However, to the extent that diffusion might reduce returns to domestic innovators, it would tend to discourage domestic innovation

[4]For recent reviews of the pertinent literature, see, for example, Wolfgang Keller, "International Technology Diffusion," *Journal of Economic Literature* XLII (September 2004): 752–782; Jonathan Temple, "The New Growth Evidence," *Journal of Economic Literature* XXXVII (March 1999): 112–156; and Gene M. Grossman and Elhanan Helpman, *Innovation and Growth in the Global Economy*, Cambridge, Mass.: MIT Press (1991); and Gene M. Grossman and Kenneth Rogoff, eds. *Handbook of International Economics*, Volume 3, Amsterdam: North Holland (1995).

and hinder growth. International technology diffusion is not automatic, because some knowledge generated in the course of R&D is tacit and cannot be codified. Countries with better educated workers and greater R&D capacity in specific fields are typically better able to absorb foreign technology and build on it in subsequent R&D activities. Thus, education and R&D capacity are important to ensure that international technology diffusion has a positive impact on domestic innovation and growth.

Countries tend to derive mutual benefit from specializing in goods and services which they trade and can produce at the lowest cost relative to other products—that is, goods and services where they have a comparative advantage. However, if one country manages to improve productivity for an export good that is the specialty of a trading partner, that partner might suffer and find itself worse off, with a resultant loss in wages and jobs.

Government funding is an important factor in R&D levels and patterns in many high-technology fields. The market in R&D services can be skewed by U.S. government support for U.S. corporations whose products have limited immediate commercial value (although they might in the long run generate important spin-offs or have dual applications) but are perceived as crucial for national security.

Some Key Trends

Some key trends in MSE R&D, discussed in more detail in other chapters of this report, are likely to determine the impact of MSE R&D globalization on the U.S. economy.

- The U.S. economy in general and MSE in particular do not stand alone. Following years of government and corporate policies aimed at opening the United States to the world market and improving access to foreign markets for U.S. firms, the economy is now much more linked to the economies of other countries.
- International technology diffusion in MSE is facilitated by the fact that many materials are already produced abroad. In recent years, the offshoring of U.S. MSE R&D has been growing along with the increasing level of global activity in MSE R&D and other types of R&D. This offshoring phenomenon, however, appears to remain relatively small—suggesting that the economic impact will be limited in the short term. For a general discussion on offshoring, see Box 5.1.
- R&D investments by foreign governments and firms, often supported by policy interventions aimed at promoting R&D in materials, are contributing to the emergence of new centers of materials R&D, such as Singapore,

> China, South Korea, and Taiwan. Some industrialized and emerging market economies have developed an educated labor force and materials production facilities (many of which moved from the United States in the 1980s), improving their ability to close the gap with the world innovation leaders and benefit from materials R&D undertaken in the leading countries. However, the ultimate effectiveness of these incentive measures is unclear.

The evidence suggests that the overall economic impact of materials R&D globalization on the United States has been limited so far, while its medium-term impact is more uncertain. In terms of R&D output, as measured by patent applications in all classes, the U.S. global leadership position remains intact. However, the margin by which it leads, as measured by patent and literature production, has narrowed. While the United States has lost competitive advantage in some materials subfields, such as catalysts, it has maintained it in others—for example, semiconductor research. The overall economic impact of globalization of materials R&D will depend on whether these trends continue and on the relative contribution of various materials subfields to the U.S. economy. In a broader sense, these trends for national security (see below) would also have a bearing on the economic impact.

Discussion

The economic impact of MSE R&D globalization is likely to differ from one subfield to another. A decline in domestic MSE R&D in selected subfields might have a negative effect on domestic growth, wages, and jobs not only in these subfields but in other MSE R&D subfields and industries dependent on materials research. However, a relative decline in one subfield might release resources to be invested in another more promising subfield where the United States enjoys a comparative advantage.

As competitive pressures from the globalization of R&D activities increase, U.S. companies should develop strategies to improve the efficiency of their domestic R&D and focus their domestic efforts on high-value, cutting-edge R&D, such as next-generation technologies and niche materials, and so on. By investing gains from cost-saving actions in new projects, the U.S. companies could generate new knowledge, products, and growth in the medium term. Relocating overseas those R&D activities in which they are relatively less efficient than their foreign counterparts could help U.S. firms to become more efficient overall and to expand domestic R&D activities. By doing so, the United States could also gain from the growing global knowledge base, which will stimulate innovation in all the leading countries. It follows that globalization could facilitate U.S. companies developing new comparative advantages in the integration of various R&D outputs, both domestic and global, into a final product. Experience in the semiconductors industry shows, for example, that the loss of U.S. dominance in DRAM memory chips was accompanied by the development of more sophisticated and lucrative microprocessors, an area in which the United States dominates.

On balance, the United States might well gain from the globalization of MSE, provided that U.S. firms and the government position themselves strategically in the new global R&D environment. The objective is to create conditions at the private and public levels whereby globalization of MSE R&D increases U.S. productivity, efficiency, and innovation capacity. U.S. companies eager to secure continued access to critical R&D and leverage domestic activities must fully integrate and take advantage of foreign R&D and international R&D relationships into their domestic R&D programs.

In summary, the global MSE R&D system is evolving rapidly, in line with broader globalization trends. The general lack of data on global flows of R&D activity and of activity in MSE R&D in particular limits analysis of economic impact. In addition, the lack of a comprehensive empirical framework for the analysis limits a robust understanding of possible impacts. Apparently, however, the impact on the relative position of the United States in MSE research has been limited so far. The medium-term impact is highly uncertain and conditional on the effective management of various risks associated with the globalization of MSE R&D at the corporate, industry, and government levels. If this can be achieved, R&D globalization can have an important positive impact on the nation's economy.

Conclusion. At this stage, economic analysis is limited by a dearth of data and by the lack of a comprehensive empirical framework. Although available evidence suggests that the globalization of MSE R&D has had a limited impact on the U.S. economy so far, the medium-term impact is highly uncertain. A positive impact will depend on globalized MSE R&D leading to increased U.S. productivity and contributing positively to U.S. domestic innovation.

NATIONAL SECURITY IMPACT

Background and Some Key Trends

After World War II and throughout the Cold War, the nation's defense services and intelligence communities were successful in their mission of protecting the United States, because in critical technologies the country maintained a one-to two-generation lead over potential adversaries. U.S. security was maintained through leadership in many fields of R&D and innovation, which together gave the nation a significant lead in military technologies.

The Hart-Rudman Commission[5] wrote as follows:[6] "The scale and nature of the ongoing revolution in science and technology, and what this implies for the quality of human capital in the 21st century, pose critical national security challenges for the United States." The commission noted that America's strength has always been tied to "the spirit and entrepreneurial energies of its people" and that "the U.S. remains the model of creativity and experimentation, inspiring other nations to recognize the true sources of power and wealth in science, technology, and higher education." It warned, however, that U.S. performance is not keeping up with its reputation and that other countries are striving hard, and, with discipline, they will outstrip America.

A number of studies over the last 5 years have considered how new threats, new adversaries, and new emerging disruptive technologies led to new challenges to which the nation and, specifically, the Departments of Defense and Homeland Security must respond. For summaries of a number of key reports, see Appendix H. The needs identified below are based on the conclusions of those reports.

It is widely accepted that the military in the 21st century will need to communicate faster, more reliably, and on a global scale. New threats require new materials for their detection. New tasks will require new weapons and new materials to make possible new and better delivery platforms. The new systems of the 21st century military will also need to demonstrate multifuctionality, self-diagnosis and self-healing, low cost, low maintenance, environmental acceptability, and high reliability. Some trends in warfare can be expected to continue: The need will increase for a precision strike force that can maneuver rapidly and effectively and survive an attack, all while distant from its command post and base. In addition, the force must be able to conceal its activities from an enemy while detecting enemy activities. Advances in information technology will increase coordination among forces, and global awareness—through real-time networked sensors and communications—will facilitate command and control and enable precision strikes. With the use of unmanned vehicles, military power will be delivered re-

[5]Officially the U.S. Commission on National Security/21st Century (USCNS/21). It was chartered in 1998 by the Secretary of Defense to carry out what was termed the most comprehensive review of American security since the National Security Act of 1947. The commission was asked to deliver a security strategy and implementation plan designed to meet the emerging challenges of the 21st century. It released three reports.

[6]*Road Map for National Security: Imperative for Change*, February 2001, http://www.nssg.gov/Reports/reports.htm.

motely and casualties will be reduced. Fighting in urban areas will increase, requiring entirely different strategies and equipment, and guerilla warfare will require new strategies and weapons.

Some high-priority military areas where it has been recommended that DOD focus its activities on several capabilities are defending against biological warfare; finding and correctly identifying difficult targets; supporting high-risk operations with systems capable of high-risk tactical operations; missile defense; affordable precision munitions that are resilient to countermeasures; enhanced human performance; rapid deployment and employment of forces globally against responsive threats; and the rapid delivery, anywhere, of "global effects." In addition, the continuing stewardship of the U.S. strategic nuclear arsenal and efforts to counteract the proliferation of nuclear materials across the globe remain a national security priority of the highest order.

Since September 11, 2001, there has been a refocusing of the nation's attention to national and homeland security. The highest priority is given to developing and utilizing robust systems for the protection, control, and accounting of nuclear weapons and special nuclear materials at their sources; ensuring the production and distribution of known treatments and preventatives for pathogens; designing, testing, and installing coherent, layered security systems for all transportation modes; protecting energy distribution services; reducing the vulnerability of ventilation systems and improving the effectiveness of air filtration in them; deploying known technologies and standards for allowing emergency responders to reliably communicate with one another; and ensuring that trusted spokespersons will be able to inform the public promptly and with technical authority whenever the technical aspects of an emergency dominate the public's concerns.

Meeting the defense needs of the country in 21st century will rely on R&D in materials and processes to improve existing materials and achieve breakthroughs in new materials and combinations. Some of the materials needed are these:

- Lightweight materials that provide functionality equivalent to that of heavier analogs,
- Materials that enhance protection and survivability,
- Stealth materials,
- Electronic and photonic materials for high-speed communications,
- Sensor and actuator materials,
- High-energy-density materials, and
- Materials that improve propulsion technology.

Future defense systems would employ advanced materials that are self-healing, that can interact independently with the local environment, and that can monitor the health of a structure or component during operation. Some advanced materials could serve as hosts for embedded sensors and integrated antennas. Others could deliver high performance in structures by protecting against corrosion, fouling, erosion, and fire; controlling fractures; and serving as fuels, lubricants, and hydraulic fluids. The next 20 years will present the materials community with daunting challenges and opportunities. Material producibility, cost, and availability requirements will be much more demanding than they are today. On the other hand, spurred by the rapid pace of advances in electronics and computation, the performance, life span, and maintainability of materials will be greatly enhanced.

Discussion

The margin of U.S. leadership in MSE R&D is eroding because knowledge and the intellectual capacity to generate new knowledge are proliferating around the globe. China, Korea, India, Japan, and the EU are making substantial investments in their R&D infrastructure, in science and engineering education, and in a variety of major R&D programs. Some nations are developing strategies for the promotion and support of innovation societies. Furthermore, U.S. companies are shifting a portion of their MSE R&D overseas as product development and technical support follow manufacturing. This process is driven by several forces, not the least of which is the substantial and growing availability of intellectual resources offshore, often at less cost, as well as by the increasing availability of unique technologies not found in the United States.

The global shifts in MSE R&D cannot be reversed or stopped. Even if the United States were to make great efforts to keep American technologies, knowledge, and capabilities under its control, the investments that other governments are making in their own domestic knowledge-creation capabilities will challenge America's military, homeland defense, and intelligence communities in their attempts to lead in technology. The loss of a national capacity in materials research, and of the manufacturing capability to take advantage of that research, is not just a matter of national pride or international image. In a knowledge-based future, only an America that continues to have access to and, in many cases, to generate cutting-edge science and technology will sustain its current world leadership in national-security and homeland-defense capabilities.

Arguably, there is a strong connection between national security and the health of the national economy. As discussed in this report, the impact of globalization on the nation's economy will depend on a reliable supply of competent knowledge workers who can produce and direct innovation in key areas. The supply of this knowledge base, however, is critical, not only for the health of the U.S. economy but also for the continued health of the country's national security.

It is possible that the globalization of MSE R&D will be like the rising tide that raises all boats—that is, increased global activity will lead to innovations, discoveries, and technologies that drive new economies and industries and open new paths for the United States to acquire access to the best materials and technologies required for national security and homeland defense. In addition, it is widely accepted that economic growth around the world, and the growth of international trade, can help underpin global security. In this sense, the globalization of research might benefit U.S. national security.

However, the benefits are not certain and neither are the risks. The following questions and suggested responses, while speculative in places, highlight important issues surrounding this uncertainty:

Will the globalization of MSE R&D reduce U.S. national capability in materials subfields of national importance?

The narrowing of U.S. leadership in certain MSE subfields is evident from the shifting trends in literature and patent statistics; flat or falling enrollments and the evolution of curricula in U.S. universities; and the loss of manufacturing and industrial-research capabilities. Some of the key trends are evident in the subfields of metallurgy—particularly in superalloys, so critical to national security—and catalysis. While the strength of these trends varies from one subfield to another, the trends themselves are clear and point to a loss in national capability in materials subfields of national importance.

Do the shift of R&D to new centers of research around the globe and the associated diminishing U.S. supply of educated workers and innovators across the spectrum of materials activity pose a risk to national security?

With low numbers of American-citizen MSE researchers, experts, and innovators being produced, research of critical national importance has become reliant on scientists and engineers trained outside the United States or on U.S.-trained noncitizens. The emergence of new centers of high-value research across the globe has led to competition for the world's best students and experts and impacts the nation's ability to attract top researchers. Any reduction in the supply of non-U.S. experts along with the acknowledged difficulty of attracting U.S. citizens to MSE will affect the overall supply of top scientists and engineers within the United States ready to conduct the MSE R&D needed for national security. Any such loss in expertise will diminish not only the value of the U.S. research output but also, in the long term, the nation's capacity to recognize, understand, and exploit the research output of the rest of the world.

What are the risks of any loss of research, innovation, human capital, or technology-deployment capability?

The risks of losses in U.S. capacity in critical areas of technology, research, and innovation and in human capital are clear. The United States has led the world in national-security-oriented R&D since the days of the Manhattan Project. A loss of leadership in U.S.-based capability will have serious implications for national security, increasing the country's dependence on technology developed (and perhaps manufactured) outside the United States and decreasing the capacity gap between the United States and the rest of the world.

What are the longer-term implications of changes in the balance of national U.S. expertise and knowledge in subfields of materials?

From a worst-case scenario perspective, would a loss in capacity in one or many MSE R&D subfields affect the ability of the United States to define what the continuing needs of the nation are? Is the capacity for the United States to continue to identify national needs in MSE at risk if U.S. experts are not being produced in sufficient numbers and with sufficient expertise? Is the ability of the United States to continue to be a leading participant in the global research effort in jeopardy? Is there a risk to the future ability of the United States to successfully analyze what the rest of the world is doing? If the answer to these questions is yes, then there is a real risk that the one- to two-generation lead the United States has enjoyed in national security technology will be eroded. However, a national effort can mitigate these risks. What is needed is a flexible strategy that recognizes that the nation's advantage may not always prevail and that adjusts accordingly.

In summary, MSE continues to be important by virtue of its contributions to the national-security and homeland-defense capabilities of the country. It is clear, therefore, that the evolution of MSE research will impact U.S. capabilities to defend against emerging threats.

Conclusion. The results of MSE R&D continue to enhance U.S. national security and homeland defense by adding improved materials capabilities to the weapons and protective systems used by today's warfighters. The evolution of materials research in the United States and abroad will affect the nation's ability not only to defend against emerging threats of the 21st century but also to ensure a healthy economy as a basic underpinning of national security. Because knowledge and the intellectual capacity to generate new knowledge are proliferating across the world, because innovation and development cycles are becoming shorter, and because U.S. dependence on foreign sources of innovation is increasing, the lead in critical technologies enjoyed thus far by the U.S. defense and intelligence communities will be seriously eroded without mitigating action.

ACCESS AND CONTROL

A key question for this study is, How can the United States ensure continued access to critical materials R&D? The nation can no longer assume that the most important MSE innovations will take place in the United States or be developed by U.S. companies in a foreign regulatory environment that allows for U.S. control of the innovations and their use.

It is easy to equate access with control. The United States has an array of regulatory environments for controlling access to information and technology related to national security. Until now, the controls on intellectual property and technologies and knowledge deemed by the federal government as critical to national security have managed to guarantee access to the results of R&D associated with this knowledge. In addition, these regulatory regimes have an impact on international cooperative R&D. Now, however, the regulations are not sensitive to the changing geopolitical environment of the 21st century and are not necessarily supportive of private-sector decisions on where to locate R&D. Faced with the globalization of MSE R&D, a one-dimensional approach to retaining access may no longer be in the national interest.

With the globalization of MSE R&D, cutting-edge innovation can be expected to emerge from both traditional centers of research—such as the United States, Japan, and Europe—and countries such as China, India, Singapore, and Korea, which have not so far been centers of important and substantive materials innovation. Many of these countries have yet to fully develop their own IP regimes to meet accepted international norms and export-control regimes for the technologies and know-how produced within their borders.

With increasing globalization, a situation can be envisioned whereby a country that has not been a traditional ally of the United States could institute control over a new and critical technology developed within its borders. Although this control might be motivated by economic rather than national security concerns, the impact could nevertheless be felt in the United States and could lose the United States a critical military or national-security advantage. In the extreme case, if continuing globalization results in the loss of a critical level of knowledge and expertise in the United States, critical innovations might go undetected and unevaluated. In the short term, the uncertainties surrounding control and ownership of knowledge and innovation might deter corporate investments in these emerging centers of R&D activity. However, in the long term the development and institutionalization of national control regimes in these countries might become an even greater concern for the United States.

Controlling and limiting access to U.S. knowledge will not, in itself, prevent the emergence of technology from new centers, and it will not guarantee the emergence of similar and equivalent innovation here in the United States, nor will it facilitate access to these new innovations. Therefore, a different approach to guaranteeing access is needed that is not based solely on control. The idea of maintaining access is at the core of the forward-looking conclusions and recommendations contained in Chapter 6 of this report.

6

Conclusions and Recommendations

Globalization of MSE R&D is defined in this study as the worldwide expansion of MSE knowledge-creation centers as a result of U.S. and non-U.S. industry and government investments along with increased worldwide collaboration facilitated by information technology. The data and evidence amassed by the committee have led to a number of conclusions, presented throughout this report and repeated in this chapter along with a summary of the rationale for each. The conclusions are focused on the status of globalization and MSE R&D and on the impacts of globalization. Also presented herein are some aggregate conclusions and recommendations concerning the question posed for this study: How can the United States ensure that it has access to current MSE R&D around the world and thereby maintain its leadership position?

OVERVIEW: SOME CONCLUSIONS ABOUT GLOBALIZATION AND ITS IMPACTS

Patent data, literature data, trends in corporate research, and the results of surveys and polls indicate increasing global activity in MSE R&D, with concomitant increases in global and transnational ownership and collaboration. MSE has become a global undertaking and is developing in a manner that is affecting U.S. leadership across MSE subfields.

Globalization and U.S. Leadership in MSE R&D

Broad trends such as the globalization of 21st century technology and increased international and transnational industrial and economic activity have helped drive the globalization of R&D. Information technology and global communications are affecting the execution of R&D in innumerable ways and enabling it to evolve new global modalities. Information from the National Science Foundation, the results of a survey of industrial R&D by the Economist Intelligence Unit, and the results of a poll of MSE practitioners carried out for this study all show that globalization has led to more transnational academia-led R&D with international academic and industrial collaborators and to more transnational corporation-led R&D with foreign affiliates of U.S. corporations, foreign academics, or foreign corporations. The data show that companies are driven to globalize their R&D activity for a number of reasons, not least of which are access to expertise, mitigating the impacts of regulatory regimes, proximity to new international customers, and cost savings. Risk factors for overseas corporate R&D investment are varied but can include concern about the ownership of intellectual property and the security of trade secrets, as well as concerns about the rule of law and democratic institutions, particularly in developing economies. Academic researchers take part in global MSE R&D by seeking out domestic or international partners that can advance their research priorities, by participating in international conferences, and by adopting information technology for sharing R&D results on a global scale.

How is globalization affecting U.S. leadership in MSE R&D? The benchmarking evidence presented in this report paints a varied picture. In the subfield of composites, the United States risks being unable to exploit the promise of composites because of the significant and continuing decline of its leadership in the subfield. Leadership in the subfield of magnetic materials is mixed, with the United States in the lead in some critical areas and among the leaders in others. The United States appears to be losing its leadership role in metallurgy R&D, and there are no signs that this trend is going to be reversed any time soon. The situation in electronic and optical-photonic materials is mixed, with the United States leading in some areas and not in others. Currently, U.S. scientists working on superconductivity are at the cutting edge in nearly all the component areas of superconducting materials. However, the United States does not dominate in any, because other countries share or surpass the U.S. lead in applications. The United States has seen a continued decline in its former dominance in catalyst technology. While the United States leads global activity in nanomaterials and nanotechnology as measured by the number of corporations engaged in the subfield, it is too early to say which, if any, region of the world is going to show clear leadership as this field matures. The use of nanotechnology in many electronic and photonic materials and devices means the U.S. position in both MSE subfields will remain interconnected.

Patent and literature surveys suggest that while MSE R&D is emerging at an accelerating rate in countries not previously known as centers of materials expertise, the United States remains either the world leader or among the world leaders across the MSE subfields. The European Union and the Asia-Pacific region, most notably Japan and most recently China, are challenging traditional U.S. leadership. Japan appears to have surpassed the United States in the alloys and ceramics subfields. Global activity in all the subfields examined during this study is diversifying, with significant increases in Asian countries that had not been active in these fields. How this trend may evolve is unclear.

In summary, it is clear that the globalization of MSE R&D is under way and is affecting U.S. leadership.

Conclusion. Globalization of MSE R&D is proceeding rapidly, in line with broader trends toward globalization. As a result of increasing international trade and investment, the emergence of new markets, and the growth of the Internet and the global communications system, MSE R&D in the United States is an internationalized activity with a diverse set of international partners.

Conclusion. The globalization of MSE R&D is narrowing the technological lead of the United States.

Impacts of the Globalization of MSE R&D

Assessing and predicting the impact on the United States of the globalization of MSE R&D is a complex, interconnected, and multidimensional exercise. The offshoring of MSE R&D from the United States and, equally important, the increasingly transnational nature of all research are raising the level and quality of materials research carried out abroad. This means that the loss of U.S. leadership in several areas of materials research where it traditionally dominated is a real possibility. The proliferation of technology and capability to new parts of the world will, at the very least, complicate the analysis of potential threats and challenges to the United States and, at worst, lead potential adversaries to gain advantage in strategic fields. While the United States might gain some advantages from exploiting the new technology that could emerge from increased global activity, the nation's strategic one- to two-generation advantage in materials-related technology and perhaps other technologies is clearly threatened.

Economic Impact

While a dearth of data and the absence of a framework for analysis limit any analysis, the economic impact of MSE R&D globalization on the United States is likely to differ across materials subfields. The evidence suggests that the overall economic impact of MSE R&D globalization has been limited so far, while its medium-term impact is more uncertain. As measured by patent applications in all classes, U.S. global leadership in R&D output remains intact. However, with increased output from other parts of the world, U.S. leadership in materials, as measured by patent and literature production, has narrowed. While the United States has lost competitive advantage in some materials subfields, such as catalysts, it has maintained it in others, such as semiconductor research. In fact, a relative decline in one subfield may release resources to be invested in another, more promising subfield where the United States enjoys a comparative advantage.

Globalization can facilitate the development by U.S. companies of new comparative advantages by integrating various R&D outputs, both domestic and global, into a final product. On balance, the United States may gain from the globalization of MSE R&D, provided U.S. firms and the government position themselves strategically in the new global R&D environment. The objective is to create conditions at the private and public levels whereby globalization of MSE R&D leads U.S. firms to increase their productivity, efficiency, and innovation capacity. U.S. companies eager to secure access to critical R&D, to leverage domestic activities, and to take advantage of foreign R&D and international R&D relationships must integrate these foreign inputs fully and effectively in their domestic R&D programs. The overall economic impact of globalization of materials R&D will depend on its evolution and on the relative contribution of various materials subfields to the U.S. economy.

Conclusion. At this stage, economic analysis is limited by a dearth of data and by the lack of a comprehensive empirical framework. Although available evidence suggests that the globalization of MSE R&D has had a limited impact on the U.S. economy so far, the medium-term impact is highly uncertain. A positive impact will depend on globalized MSE R&D leading to increased U.S. productivity and contributing positively to U.S. domestic innovation.

National Security Impact

U.S. security forces in the 21st century will need to communicate faster, more reliably, and on a global scale. New threats to national and homeland security will require new materials for their detection. New tasks will require new weapons and new materials to make possible new and better delivery platforms. The new systems will also need to demonstrate multifuctionality, self-diagnosis and self-healing, low cost, low maintenance, environmental acceptability, and high reliability. Meeting these needs will depend on R&D in materials and processes to improve existing materials and achieve breakthroughs in new materials and combinations.

With the global shift in MSE R&D activity, even if the United States makes great efforts to keep American technologies, knowledge, and capabilities under its control, the investments of other governments in their own domestic knowledge-creation capabilities will challenge the technology lead of America's military, homeland defense, and intelligence communities.

Conclusion. The results of MSE R&D continue to enhance U.S. national security and homeland defense by adding improved materials capabilities to the weapons and protective systems used by today's warfighters. The evolution of materials research in the United States and abroad will affect the nation's ability not only to defend against emerging threats of the 21st century but also to ensure a healthy economy as a basic underpinning of national security. Because knowledge and the intellectual capacity to generate new knowledge are proliferating across the world, because innovation and development cycles are becoming shorter, and because U.S. dependence on foreign sources of innovation is increasing, the lead in critical technologies enjoyed thus far by the U.S. defense and intelligence communities will be seriously eroded without mitigating action.

MAINTAINING ACCESS

Numerous policy reports over the years have found that MSE R&D remains in the national interest and recommended that the United States should remain among the world leaders across all MSE subfields and the leader in many. The present study concludes that the globalization of MSE R&D is under way and is affecting U.S. leadership across MSE subfields. The impact of materials R&D globalization can be positive and large, but the risks of a negative impact remain substantial. To avoid a negative impact, the U.S. government and private sector must exploit foreign or joint R&D to benefit domestic innovation by integrating it efficiently and effectively into domestic civilian and military R&D.

With the increasingly global nature of MSE R&D the question that arises is how the United States can maintain access to the global output of MSE R&D and thereby maintain a leadership position. Access is only one part of the story. Even if there is access, it might not be full access or access to the right sort of technology. Integration also must be a priority, but integrating R&D is not easy. There are risks, including that some knowledge generated by foreign R&D will not be absorbed in the United States and that there may not be sufficient domestic U.S. expertise to recognize the foreign innovation and maximize its integration. Maintaining access to current MSE R&D will require active management so as to mitigate the potentially negative economic and national security impacts of globalization. Such active management requires government action.

Conclusion. In response to the globalization of MSE R&D, it is the task of public policy to minimize the risks and maximize the benefits to ensure the ongoing U.S. innovation that is essential to the nation's economy and national security and to facilitate continued access to the new knowledge generated by MSE R&D.

RECOMMENDATION ON DEVELOPING A NATIONAL STRATEGY

To maximize the benefits for the United States of the globalization of materials science and engineering research and development (MSE R&D), the federal government should create a well-defined and coordinated national strategy to manage the development of and access to strategic MSE knowledge and technology in a global framework. Particular emphasis must be given to defining and achieving MSE R&D goals for ensuring a strong 21st century U.S. military and a secure U.S. homeland.

In building a U.S. national strategy for effective development and use of MSE R&D, the following elements should be considered:

- Identifying in MSE R&D across the defense services and other relevant national security agencies programmatic linkages that will facilitate a coordinated approach to answering critical questions across the subfields of MSE and assessing the readiness of R&D programs to do so, analyzing domestic readiness to provide critical MSE capabilities, and developing recommendations on the role that international and transnational MSE R&D might play;
- Defining (1) immediate priorities for which programmatic directions are clear and (2) next steps, which will require development of a roadmap as a prelude to determining relevant MSE R&D programs;
- Including as participants a comprehensive range of stakeholders and decision makers from the defense, homeland security, and intelligence communities and obtaining significant input from and coordinating with the wider federal science and engineering agencies—including the National Science Foundation, the Department of Energy, NASA, and so on; and

- Soliciting independent advice from academia, industry, and other experts, as required—perhaps with the participation of the Defense Science Board—and obtaining input from industry regarding policies and incentives that could encourage proactive industry strategies for sustaining a strong MSE R&D base in the United States.

Building a robust and effective national strategy for ensuring U.S. access to the results of MSE R&D will also require obtaining a better understanding of current trends in MSE R&D worldwide; defining in a clear and focused manner the critical questions and challenges MSE R&D must address to meet national economic, defense, and homeland security needs; and developing a renewed approach to managing regulatory regimes, improving the education system, and strengthening the infrastructure for U.S. MSE R&D. With that in mind, the following immediate recommendations are made.

Maintaining Access with Better Monitoring

It became clear during the course of this study that there are few data on the flow of investments in R&D generally and in MSE R&D specifically. The discussion of how the nation should react to the globalization of MSE R&D will need to be informed by better data. In addition, data alone will not be sufficient, because new analytical tools are needed that reflect the complexity of R&D globalization.

RECOMMENDATION ON GATHERING BETTER DATA

U.S. data collection efforts and forecasting systems should be strengthened in order to monitor trends in the offshoring of MSE R&D and the growth of MSE R&D worldwide.

Addressing the national need for better data and monitoring will require collective efforts on the part of the Department of Defense, the Department of Commerce, and other data-collecting and intelligence agencies. Ideas to be considered include periodic surveys of corporations on their materials R&D. Obtaining meaningful data on materials is difficult, because materials are intermediate products of other products. In a sense, materials are silent (in terms of data), so that trends can be easily obscured in the more general statistics.

Current but limited DOD information-gathering activities around the world could be broadened and improved to assure access to international innovation. Developing the tools to maintain access to global R&D requires consideration of how DOD internal technology forecasting systems could be strengthened and how current DOD initiatives to identify critical technology around the world could be expanded. New technology could facilitate the identification of synergistic needs across the military services and the homeland-defense services, helping reduce further any stovepiping of R&D within DOD. Technology should allow the development of clearinghouses for R&D goals and needs, initially by subfield and perhaps more broadly in the future.

RECOMMENDATION ON IMPROVING MONITORING

The Department of Defense should build on existing capacities to monitor, assess, and promote access to developments in MSE R&D across the globe with a strategic view to underpinning the maintenance of U.S. leadership and security. In addition, existing U.S. government internal systems for strategic and critical technology analysis, management, and integration should be strengthened. Modern database and communication systems for identifying synergies across the defense services should be developed.

An international DOD-led program could think globally about MSE R&D infusion into the armed forces through liaison visits, visitor support programs, conference support programs, and the promotion of cooperative and collaborative programs. The enhanced activity could provide DOD and the services with a window on global MSE R&D innovation and avenues for maintaining access. To remain current and active in a globalized MSE R&D theater, the United States must adopt an international and focused initiative to infuse international and domestic MSE R&D into the armed forces of the 21st century.

Maintaining Access with Better Benchmarking

The global environment for MSE R&D is changing rapidly and the U.S. relative position in many subfields of MSE is in a state of flux. Any strategy to respond to globalization must have at its foundation not only better monitoring of global activity but also a thorough knowledge of the relative status of U.S. MSE R&D. The benchmarking carried out for this study was limited owing to the breadth of the overall study objectives and the constraints on time and resources. A more comprehensive benchmarking study is needed. Such a study would have to be carried out in expedited fashion and soon, because the U.S. competitive position in MSE R&D is changing rapidly and could be dramatically eroded in some subfields without corrective action. Such a study could also carry out longitudinal comparisons of and interconnections between MSE subfields and compare the impact of globalization on MSE R&D and R&D in other engineering fields.

RECOMMENDATION ON CONDUCTING COMPREHENSIVE, EXPEDITED BENCHMARKING

An expedited benchmarking study, similar to Experiments in International Benchmarking of U.S. Research Fields *(National Academy Press, Washington, D.C., 2000), should be conducted immediately to assess the relative global position of the United States in MSE R&D.*

Maintaining Access with a Better Understanding of Long-Term Security Needs

Building a successful strategy to respond to globalization requires a thorough understanding of the knowledge needs of the national security and homeland defense systems. Defining those needs can complement analyses of more immediate military needs by taking a longer-term perspective on the investment strategy required for establishing a military for the 21st century and securing the homeland. "Longer-term" means taking a two-decade and at times speculative perspective. Earlier reports by the National Academies, the Defense Science Board, and others could feed into the development of a clear list of questions that need to be answered over the long term and challenges that need to be met to maintain the nation's lead in technology for national and homeland security.

RECOMMENDATION ON ESTABLISHING LONG-TERM SECURITY NEEDS AND CHALLENGES

The Department of Defense should strengthen current systems for establishing clearly the materials needs of the 21st century warfighter as well as those essential to achieving national and homeland security priorities. Efforts in this regard should focus not on meeting the shorter-term acquisition needs of the military, but rather on identifying and prioritizing the longer-term questions and challenges that MSE R&D will have to address in order to meet identified long-term U.S. security needs.

The potentially adverse effects of globalization on traditional U.S. leadership in R&D and on U.S. capacities and capabilities are real and present. In assessing the national security impacts, complacency is not a policy option. An analysis premised on the shorter-term and near-future technological requirements of the warfighter and on a survey of the supply of current technologies on the world market risks being myopic. The national security of the country is more interconnected and multidimensional than an analysis of immediate or predicted acquisition needs of the military would suggest. Neither innovation nor future threats can always be foreseen and predicted. It may not always be clear today what a one- or two-generation lead over potential adversaries tomorrow might require in terms of new materials and technologies. In addition, previous success in acquiring required technology from global sources does not guarantee future supply. Less tangible and harder-to-predict threats may also arise from the increasingly global nature of MSE R&D and, in particular, from any concomitant loss of U.S. expertise or the proliferation of new technology to terrorist organizations or adversarial regimes.

Existing systems within the government for assessing technologies critical to the 21st century warfighter and to wider national security concerns will benefit from the highest level of coordination and cooperation within DOD and between the relevant federal agencies. Continued review facilitates an assessment of existing critical technology lists, contractual arrangements, R&D funding procedures, and so on and helps in defining goals and challenges for MSE R&D.

Maintaining Access with Better Management of Regulatory Regimes

The rapid evolution of the environment for MSE R&D and the need to develop a national response thereto requires consideration of how the nation's regulatory regime needs to react to the realities of R&D in the 21st century. As shown in this report, regulatory regimes can be a driver for decisions on locating R&D. It seems clear that questions about the security of intellectual property developed abroad—whether the export license process will hinder the effective execution of an R&D program, whether workers will be available to carry out the research, and what tax incentives might be in place in the locales under consideration—would have to be addressed in any thorough decision-making process. It is unlikely, however, that any one of these issues would be the sole determinant of the location of an R&D activity. Nevertheless, it is incumbent on makers of public policy to ensure that as R&D becomes a more global activity, U.S. researchers are not unreasonably impeded by U.S. regulatory regimes from taking part in international activity of national importance. Continuing review should build on current management and review systems and consider if and how existing frameworks need to adapt in a systematic fashion to the realities of the global R&D theater.

RECOMMENDATION ON REVIEWING REGULATORY REGIMES

A systematic review of the rationale for and the impacts of U.S. government regulation of the transfer of knowledge and innovation across borders within the framework of globalized MSE R&D should be carried out by a government task force of representatives from the relevant agencies, with input from academia and industry.

The review could seek input from both the defense and commercial sectors of the private sector and will also assess and build on existing internal government systems designed to maximize the effectiveness of the existing critical technology lists, export regimes, and so on. The review should ask whether there are technologies the nation need not necessarily control but to which it merely needs secure access. The next question would be how such a change in intent would affect the nation's control regime. In addition, the review might consider innovative approaches. For instance, a risk management approach could be adopted in which defense contractors and researchers (1) identify the key risks of their global R&D strategies for identified critical technologies and (2) describe the risk-management systems they have developed to ensure continued access to these technologies. This approach would have the advantage of being private-sector driven: While allowing the government to influence industry decisions by bringing public, longer-term national security considerations to the table, it would also allow firms to organize their global R&D activities flexibly.

The intent of the review proposed here is not to create new technology lists but to reevaluate the government's systems for critical technology and knowledge management. The idea would be to ensure access to all the relevant technologies without overburdening private firms and other researchers with regulations and without stunting U.S.-led innovation in the global MSE R&D theater.

Maintaining Access by Remaining on the World Stage

There may be a temptation to protect U.S. interests by retreating from the world stage in areas deemed critical to national security or to the economic interests of the country. Calling into question the exemption from ITAR of basic research would be one such protectionist move. An isolationist approach, however, would have significant costs because it would require a large government investment to maintain a sufficiently robust and productive capability in critical areas. More important, retreating into such a protectionist posture might deprive the United States of superior technologies developed outside the United States and, of most concern, those developed in countries that are not the closest military allies of the United States. Access to cutting-edge knowledge and technology can be better and more effectively guaranteed by the United States being the most active player in the global MSE R&D effort.

Conclusion. It is in the long-term interest of the United States to participate in international partnerships in MSE R&D and thereby ensure U.S. access to cutting-edge knowledge and technology.

Notwithstanding the imperative to remain an active participant in international R&D, a number of other issues arise in this context that will need to be addressed by policy makers. For instance, how will the U.S. balance the risks and benefits of taking part in global MSE R&D, particularly in materials and technologies critical for the nation's defense? How much domestic expertise and activity in key areas of MSE R&D does the country need to keep on the back burner in case the non-U.S. supply is cut off? How might the country choose which subfields are worth preserving in the United States? Which technologies and expertise must the country retain control of, and which does it simply need continued access to? These are the kinds of questions that the discussion leading to a national strategy will have to answer.

Maintaining Access by Managing the Education System and Maintaining an Infrastructure

Maintaining a domestic U.S. capacity to engage in international MSE R&D, to integrate non-U.S. R&D into U.S. systems, and to monitor and understand global MSE R&D and its impact on U.S. leadership will require the U.S. educational system to produce the MSE practitioners necessary to achieve these goals. As shown in this report, that system faces several challenges in this regard, including the increasingly broad curricula in MSE departments; the decreasing attractiveness of MSE as a career choice for high school and university graduates; and the continuing dependence of graduate programs on foreign students, who must be attracted to the United States in an increasingly competitive global market for the best students. It is not clear that the current MSE education system, including university research, is producing graduates with the depth of knowledge to meet the nation's needs.

Because the health of the country's MSE education system is of direct interest to the federal agencies that support MSE activities, such as DOD, NSF, and DOE, as well as to the materials industry, these agencies should cooperate to spell out the national educational needs in the field. Industry roadmaps identifying industrial needs could also be part of the effort to guide MSE education in the decades ahead.

> **Conclusion.** The MSE education system, including K–12 mathematics and science education, will have to evolve and adapt so as to ensure a supply of MSE professionals educated to meet U.S. national needs for MSE expertise and to compete on the global MSE R&D stage. The evolution of the U.S. education system will have to take into account the materials needs identified by the federal agencies that support MSE R&D as well the needs of the materials industry.

A number of issues could be considered in examining the education system: the promotion of MSE as a career choice; overcoming the deficiencies of students entering MSE programs from the K–12 system; meeting the evolving needs of U.S. industry and other actors; the possibility of targeted financial support to encourage U.S. citizens to study materials; the possibility of setting minimum competences for graduate students working on master's degrees in MSE; the role of smaller MSE educational programs, the correct balance between large and small departments, and the role of accreditation; and the best strategy to produce graduates competent in all four aspects of materials, including materials processing, an area in which the United States was already behind the rest of the world, according to the 2000 benchmarking report of COSEPUP.[1] The key issue is how to produce the best possible corps of MSE graduates educated to meet the nation's needs and to compete in the global MSE R&D theater.

Maintaining expertise and leadership will be based on a robust research infrastructure. The data presented in this report clearly illustrate the huge technical downsizing of the U.S. metals industry in the last few decades of the 20th century. As stated repeatedly herein, U.S. leadership is being challenged in the subfields of MSE. The well-documented reductions at once-dominant industrial materials laboratories, exemplified by the loss of much of the effort at Bell Laboratories, indicate the broad range of materials-intensive industries where the R&D infrastructure is being lost. A unique attribute of industrial R&D is that process methodology is not a final step, as often is the case in academic research. Rather, the development of a process methodology is a first step in the definition of a viable industrial route to production at the industrial scale. The function of this kind of industrial activity is to explore new technical areas that could be of business interest to the company, maintain an awareness of what is happening in university or government laboratories, and transmit these findings to the relevant business units. It remains unclear as to where such work will take place in the future as this industrial element of the national R&D infrastructure is eroded. How will science be brought to market?

Conclusion. There is a need to maintain a robust U.S. MSE R&D infrastructure whereby materials problems can be addressed and solved and the solutions verified, from laboratory through pilot scale.

[1]National Academy of Science, National Academy of Engineering, Institute of Medicine, *Experiments in International Benchmarking of U.S. Research Fields*, Committee on Science and Engineering Public Policy, Washington, D.C.: National Academy Press (2000).

FINAL REMARKS

Addressing the needs that became clear during the course of this study, mitigating the risks identified, and answering outstanding questions—most of which boil down to matters of risk tolerance and management—require the nation's public policy makers to formulate a national response strategy to the globalization of MSE R&D. The recommendations in this chapter provide a framework for a robust strategy that will assure a positive impact and outcome for the United States and the nation's continued access to current MSE R&D. The framework is based on a series of initiatives that will benchmark MSE R&D in the United States, define the MSE R&D challenges and opportunities in meeting 21st century national security needs, manage an IP regulatory framework that supports U.S. MSE innovation in a globalized environment, and build a national infrastructure to support a global role for the United States. The challenge here is multidimensional and intrinsically interconnected across many agencies within the federal government.

In closing it is worth noting that the President's mandate for defense transformation is to "challenge the status quo and envision a new architecture of American defense for decades to come." In addition, the White House's Office of Management and Budget has identified R&D for homeland and national security as a presidential priority. Number five of the top five priorities of the DOD Office of Force Transformation is to "discover, create, or cause to be created new military capabilities to broaden the capabilities base and to mitigate risk." Achieving these goals will require each sector—that is, government, industry, and academia—to actively manage by means of a well-defined strategy its participation in the increasingly global research and development theater.

Appendixes

A

Committee Biographies

Peter Bridenbaugh, *Chair*, retired in 1998 from Alcoa as an executive vice president responsible for expanding the use of aluminum in automobiles and integrating Alcoa's technical and commercial initiatives in the automotive market. He joined Alcoa in 1968 at Alcoa Laboratories, and after a number of positions was appointed executive vice president and chief technical officer in 1991, at which time he was given overall responsibility for R&D, engineering, and environment/safety/health. In 1994, he was given direct responsibility for the automotive market. He led Alcoa Laboratories from 1983 until May 1995. Dr. Bridenbaugh received a bachelor's degree in mechanical engineering and a master's degree in metallurgy from Lehigh University and a Ph.D. in materials science from the Massachusetts Institute of Technology (MIT).

Dr. Bridenbaugh has served on advisory boards for several universities throughout the United States and two government laboratories. He was a double subject editor—corrosion and nonferrous metals—for the *Encyclopedia of Materials: Science and Technology* and served on the advisory committee for writing the history of Corning, Inc. He has chaired national conferences for the Federation of Materials Societies and the Industrial Research Institute (IRI). Dr. Bridenbaugh also served on several corporate boards, including Precision Castparts Corporation and Keystone Powdered Metal Company. He is a member of the National Academy of Engineering, Sigma Xi, the American Institute of Mining, Metallurgical, and Petroleum Engineers (AIME), the American Society for Metals (ASM), The Minerals, Metals and Materials Society (TMS), the Materials Research Society

(MRS), and IRI. He has received the following honors: National Materials Advancement Award, Federation of Materials Societies; Hoyt Lecture, American Foundryman's Society; fellow, ASM International; Zae Jeffries Lecture, ASM; Leadership Award, TMS; Alpha Sigma Mu Lecture, ASM-TMS; Andrew Carnegie Lecture, ASM; Distinguished Lecture on Materials and Society, ASM-TMS and ASM Honorary Membership-2004.

Miller Adams is vice president of Boeing Technology Ventures, a unit of Boeing Technology and Boeing Phantom Works, the research and development organization of the Boeing Company. He leads a team responsible for the overall technology planning process for Boeing. He also is responsible for certain aspects of external technology acquisition strategies for Boeing, including the evaluation of external technology solutions, international industrial technology programs, strategic technology alliances, global university research collaborations, venture capital investments, and Boeing's overall global R&D strategy. Mr. Adams also is responsible for Boeing's internal incubator program, known as the Chairman's Innovation Initiative, and for value-creating strategies around spin-in business opportunities built on Boeing technologies. He received a bachelor of arts degree from Seattle University and a law degree from the University of Puget Sound (now Seattle University School of Law).

Ashish Arora is a professor of economics and public policy at the Heinz School of Public Policy and Management at Carnegie Mellon University. Dr. Arora's research centers on the economics of technological change, the management of technology, intellectual property rights, and technology licensing. He has worked on the productivity of university research and the growth and development of biotechnology and the chemical industry. He is a codirector of the software industry center at Carnegie Mellon University and is studying the development of the Indian software industry and its links to the United States. Dr. Arora has published over 50 articles and coauthored two books on technology and the market economy. He is on the editorial boards of the *Journal of Management and Governance* and *Research Policy.* Professor Arora earned his Ph.D. in economics from Stanford University in 1992.

Gilbert Benavides is a Distinguished Member of the Technical Staff in the Manufacturing Science and Technology Center at Sandia National Laboratories. He has a B.S. from the University of New Mexico and an M.S. from Stanford University, both in mechanical engineering. He has been the project leader for many R&D projects that have led to the development of new devices and novel advanced manufacturing processes. His research areas include developing and testing new mesomachining manufacturing processes, such as focused-ion-beam machining, machining with microtools, femtosecond laser machining, and microelectrodischarge machining. The mesomachining work won a best paper award at the 2000 Defense Manufacturing Conference. He led a project to develop a multi-degree-of-freedom silicon-based MEMS device. He just completed a project to develop a manufacture method to package electromicrofluidic devices and is developing a drop ejection system capable of patterning micron-size drops of fluids onto a substrate. Other technical development projects have led to new devices, including a hydraulic control system for reentry vehicles, an electromechanical nuclear safety component, and a down-hole seismic imaging system.

Uma Chowdhry is vice president for Central Research & Development (CR&D) at DuPont. She joined DuPont in 1977 as a research scientist in CR&D at the DuPont Experimental Station. Dr. Chowdhry was elected a fellow of the American Ceramic Society in 1989. She has served on the advisory boards of engineering schools at MIT, the University of Pennsylvania, Princeton University, and the University of Delaware, as well as on the program advisory board and election subcommittee for the National Academy of Engineering. She was recently elected to the board of directors for the IRI, to the national Inventors' Hall of Fame, and to a laboratory operations board for the U.S. Department of Energy. Dr. Chowdhry has been a member of the National Committee on Women in Science and Engineering of the National Academy of Science and the National Academy of Engineering since 1999. Born and raised in Mumbai, India, she came to the United States in 1968 with a B.S. in physics from the Indian Institute of Science, received an M.S. from Caltech in engineering science in 1970, and a Ph.D. in materials science from MIT in 1976.

Edward Dowling is group director, Group Mining and Exploration for DeBeers Group Services. In this capacity he is responsible for all technical aspects of DeBeers investments in global diamond mining partnership operations, global exploration, and ongoing technological development. Prior to joining DeBeers, Dr. Dowling was executive vice president for operations with Cleveland-Cliffs, Inc., in charge of the largest North American iron producer. He has profit-and-loss responsibility for six large iron ore mining, processing, and manufacturing operations; an international reduced iron facility that employs a novel fluidized-bed reactor technology; and overall company-wide improvement efforts. Dr. Dowling has held a progression of technical and operating positions throughout his career. Prior to joining Cleveland-Cliffs, he was senior vice president and director of process management and engineering with Cyprus Amax Minerals Company, at that time the largest U.S.-based mining enterprise. While with Cyprus, he also led its subsidiaries Climax Molybdenum Company and Climax Specialty Metals and Performance Chemicals, as well as downstream copper smelting and refining operations. He is recognized in the industry for his process engineering expertise and the ability to integrate engineering theory and practice to obtain real solutions to important industrial problems. He holds a B.S. degree in mining engineering and M.S. and Ph.D. degrees in mineral processing, all from the Pennsylvania State University. For his contributions, Dr. Dowling has received a number of industry awards, most recently from the Extractive Processing Division of TMS (2000) and the Robert H. Richards Award from AIME (2001). He is a member of TMS, Sigma Xi, the Mining and Metallurgical Society of America, the American Iron and Steel Institute, and others. He has published more than 35 articles with an emphasis on processing engineering and technical approaches to operations and business optimization.

Gordon Geiger earned his B.E. degree in metallurgy at Yale University and his M.S. and Ph.D. in metallurgy and materials science at Northwestern University. In 1973, he joined the University of Arizona as professor and head of the Department of Metallurgical Engineering. During his teaching career, he trained over 50 graduate students, collaborated in the writing of several well-known textbooks, and consulted for industry and government. After 15 years of teaching and research, Dr. Geiger returned to industry in 1980 as an internal consultant at Inland Steel and later as a vice president of Chase Manhattan Bank. From 1983 to 1993, he served as vice president and executive vice president of Cargill's North Star Steel company. He left North Star and Cargill in 1993 to start a new company, Qualitech Steel. He now lives in Tucson, Arizona, where he consults and where he has established a B.S. program in engineering management at the University of Arizona. He is director of the program and professor of industrial and systems engineering. He is the recipient of the A.B. Campbell Award from the National Association of Corrosion Engineers, the Bradley Stoughton Award from ASM, the Leadership Award from TMS-AIME, the Charles W. Briggs Award and the Robert Woolson Hunt Award from the Iron and Steel Society (ISS) and AIME, and was a member of the U.S. delegation to the First U.S.-People's Republic of China Metallurgical Conference. He gave the Howe Memorial Lecture for ISS-AIME in 1999 and was an Alpha Sigma Mu lecturer for ASM. He served as president of TMS-AIME, as president of the Accreditation Board for Engineering and Technology, and as president of ASM International.

Jennie Hwang is president of H-Technologies Group and CEO of Asahi America, Inc. She has held senior executive positions with Lockheed Martin Corp., Sherwin Williams Co., SCM Corporation, and International Electronic Materials Corporation. She is also an invited distinguished adjunct professor in materials science and engineering at Case Western Reserve University. Dr. Hwang's career encompasses research management, technology transfer, international business, corporate executive positions, CEO of three start-up companies, corporate and university governance, and presidency of professional organizations. Among her many awards and honors are citations by the U.S. Congress and the Ohio Senate/House for outstanding achievement; membership in the National Academy of Engineering (1997); induction into the Women in Technology International Hall of Fame; being named a "Star to Watch" by *Industry Week* magazine; and induction into the Ohio Women's Hall of Fame. She is internationally recognized as a pioneer in the miniaturization and manufacture of electronic devices and infrastructure development, particularly in materials and processes. She is the inventor of a number of patents and the author of 250 publications, including sole authorship of several internationally used textbooks. As a columnist for *SMT*, a globally circulated trade magazine, she addresses global market thrusts and technology issues monthly. In addition, she is a prolific author and speaker on education, workforce, social, and business issues; emerging technologies at the U.S. Patent and Trademark Office; and the dissemination of new technologies. She has served as an advisor to university, government, and industry. She also serves on the board of Fortune 500 NYSE-traded companies and the civic and university boards. Her formal education includes two M.S. degrees, one in chemistry and one in liquid crystal science from Columbia University and Kent State University, respectively, and a Ph.D. in materials science and engineering from Case Western Reserve University, where she serves as a board director.

Michael Jaffe is a research professor with the New Jersey Institute of Technology in the Biomedical Engineering Department. He is also chief scientist for industrial programs and director of the Medical Device Concept Laboratory in the New Jersey Center for Biomaterials and an associate research professor at Rutgers University. His expertise is in innovative materials research, such as biomimetics, and in DOD system applications. Dr. Jaffe's work has focused on understanding the structure-property relationships of polymers and related materials, the application of biological paradigms to materials design, and translation of new technologies to commercial reality. He was the recipient of the 1995 Thomas Alva Edison Patent Award, presented by the R&D Council of New Jersey. He is a fellow of the American Association for the Advancement of Science (AAAS) and the U.S. National Committee for the International Union of Pure and Applied Chemistry.

Robert Pfahl received his Ph.D. in mechanical engineering from Cornell University, where he majored in heat transfer and fluid mechanics. He is the vice president of operations for the International Electronics Manufacturing Initiative (iNEMI). Before this, he was the director of international and environmental research and development at Motorola's Advanced Technology Center, where he led Motorola's environmental technology R&D. He was also responsible for globalizing Motorola's Advanced Technology Center, including research centers in China and Germany. Dr. Pfahl is a member of the steering committee of the International Society of Industrial Ecology. He is also a senior member of the Institute of Electrical and Electronics Engineers (IEEE) and a member of the International Microelectronics and Packaging Society (IMAPS). He holds nine U.S. patents in electronics manufacturing technology and is the inventor of the vapor-phase soldering process. He led the U.S. electronics industry in its preparation of the 1994 and 1996 National Electronic Manufacturing Initiative Roadmaps. Dr. Pfahl chairs the National Roadmap Coordinating Committee, which coordinates U.S. electronic roadmapping activities with U.S. government activities in electronics R&D. In recognition of his efforts to eliminate the use of chlorofluorocarbons (CFCs) in the electronics industry, Dr. Pfahl received the U.S. Environmental Protection Agency's Stratospheric Ozone Protection Award for "executive leadership and industry organizing" in 1991. Dr. Pfahl also chaired the American Electronics Association's CFC task force.

Natalia Tamirisa is a senior economist with the International Monetary Fund, where she has worked on a range of topics in international economic policy, focusing on emerging market economies. As the holder of an M.Sc. in aerospace economics from the Moscow Aviation Institute, in Russia, Dr. Tamirisa also has a technical background in the economics and management of aerospace research and development. She received honors, scholarships, and grants for her graduate studies and an outstanding research award for her Ph.D. thesis. She was the valedictorian at the International Space University summer session in Toulouse, France, in 1991. Dr. Tamirisa has authored and coauthored publications on issues in the globalization of trade and finance, including trade and capital controls, dual-use technologies, intellectual property rights, and the environmental economics of space.

Xishan Xie is a professor in the High Temperature Materials Testing and Research Laboratories at the University of Science and Technology in Beijing. He is also vice chairman of the International Affairs Committee of the Chinese Society for Metals. Dr. Xie also serves as the Chinese delegate to the International Organization of Materials, Metals, and Minerals Societies. He has participated in several Pacific Rim International Conferences on Advanced Materials and Processing. He has published nearly 200 papers in English and Chinese versions for technical journals and proceedings.

B

Presentations to the Committee at its Public Meetings

Meeting 1, January 26, 2004

Corporate perspective on international R&D
  Bob Pfahl, International Electronics Manufacturing Initiative
Department of Commerce view
  Les Smith, National Institute of Standards and Technology
Drivers for international R&D
  Ed Dowling, Cleveland-Cliffs, Inc.[1]
Materials research and education in the United States and abroad
  Merrilea Mayo, Government-University-Industry Research Roundtable

Meeting 2, March 14, 2004

Discussion of meeting with the Bureau of Industry and Security at the Department of Commerce
  Toni Maréchaux, NRC
  Natalia Tamirisa, International Monetary Fund
Professionals without borders
  Dan Thomas, The Minerals, Metals and Materials Society

[1]Ed Dowling has since moved to the DeBeers Company.

DOD approaches to international research
Phil Parrish, University of Virginia
Introduction to the forum
Peter Bridenbaugh
Materials research in China
Xishan Xie, University of Science and Technology, Beijing
Materials research in Japan and Korea
Emily Ann Meyer, NRC
Materials research in India
Michelle Iacoletti, NRC intern
Materials research in the European Union
Duane Shelton, World Technology Education Center
Materials research in the former Soviet Union
Bert Westwood, Sandia National Laboratories
Discussion of the bottom line
Natalia Tamirisa, International Monetary Fund

Meeting 3, May 12–13, 2004

Global S&T economic issues
Bernard Kritzer and Alex Lopes, Department of Commerce
Export compliance: A corporate perspective
Mark Snyder, The Boeing Company
Small business R&D: A corporate perspective
Anthony C. Mulligan, Advanced Ceramics Research
Policies concerns and intellectual property
Tim Holbrook, University of Chicago-Kent College of Law
Tax policy and global decision making
Jessica L. Katz, Caplin and Drysdale, Attorneys
ITAR fact-finding results
Toni Maréchaux, NRC
Protecting the researcher: A university perspective
Robert Anderson, Illinois Institute of Technology

Meeting 4, July 22–23, 2004

Overview of previous benchmark study
Kathy Taylor, General Motors (retired)

C

Global Trends in MSE Patents

ALL PATENTS

Before looking at the indicators for some MSE subfields, looking at the global trends in patenting will help identify baseline trends against which MSE-specific trends can be compared. The number of patents awarded to inventors located in the United States and in 10 other key countries—France, Germany, Italy, Switzerland, and the United Kingdom formed the Euro5 group and China, India, Japan, South Korea, and Taiwan, the Asia5 Group—was measured over a 25-year period in 5-year increments. The data in Figures C.1 and C.2, from the U.S. Patent and Trademark Office (USPTO), show that the United States dominates the overall share of U.S. patents awarded, with its total share of 53 percent outstripping the combined share of the Euro5 Group and Japan. However, caution is warranted and advisable because of what is called the home advantage. The Organisation for Economic Co-operation and Development's (OECD's) *Compendium of Patent Statistics* reports that data drawn from a single patent office display a bias whereby proportionate to their inventive activity, domestic applicants are more likely to file for patents in their home country than foreign inventors (see Figure C.3).

The OECD has developed a set of indicators based on a set of patents taken at the European Patent Office (EPO), the Japanese Patent Office (JPO), and the USPTO. Using these indicators eliminates the home advantage while also eliminating lower-value patents, which are unlikely to be patented in all three locations. An analysis of the geographic distribution of these indicators is shown in Figure

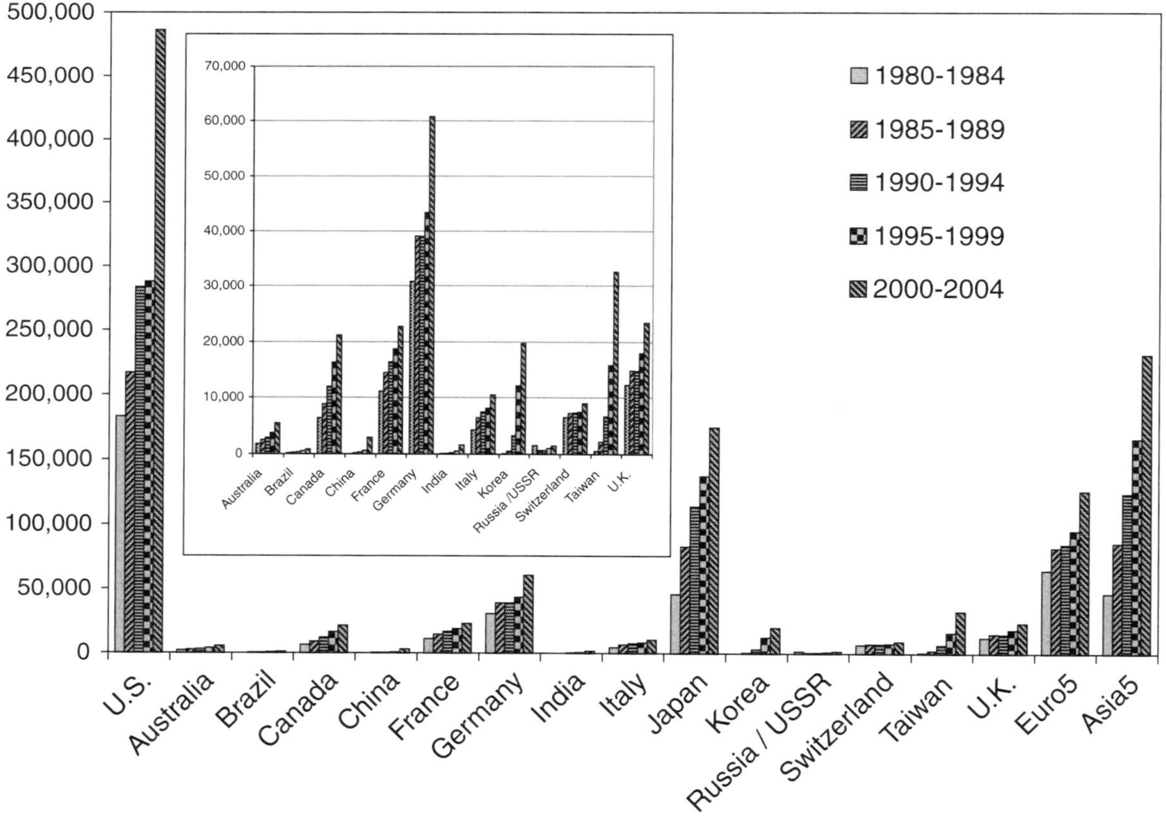

FIGURE C.1 Number of U.S. patents in all classes awarded between 1980 and 2004 (up to October 21, 2004). The Euro5 data combine the data for the five West European countries, and the Asia5 data combine the data for the five Asian countries. The insert shows the data without the data from Japan, the United States, and the Euro5 and Asia5 groups. SOURCE: USPTO.

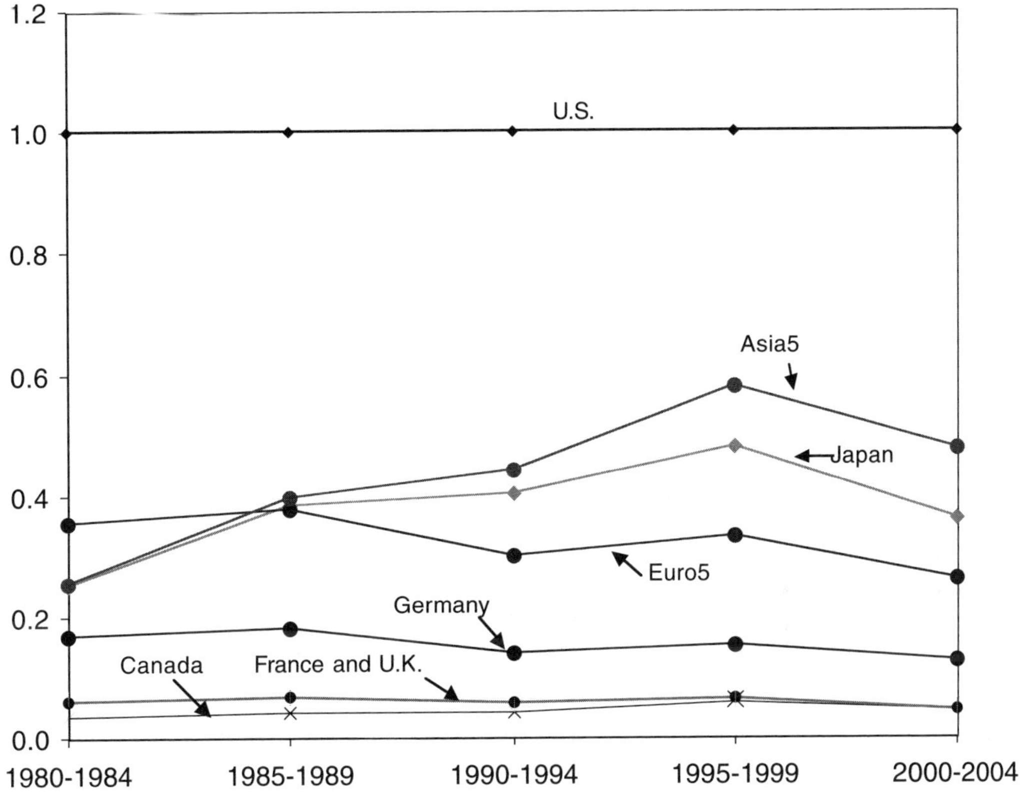

FIGURE C.2 Patent data for all classes normalized to U.S. numbers to show relative activity. The United States remains dominant over time. The divergence of the Asia5 and Japan data lines show emerging patent activity in Taiwan, Korea, China, and India. The dip in share for Asia and Europe in the last 5-year period is linked to a jump in U.S. patent activity over the same period, perhaps linked to the high-tech economic boom of the late 1990s. SOURCE: USPTO.

C.4, which shows the ranking in terms of share. The United States, Japan, and the European Union dominate the ranking. The U.S. and EU shares did not change significantly between 1991 and 2000, while the Japanese share decreased. Countries often associated with the globalization of R&D—such as China, India, Korea, and Taiwan ("Chinese Taipei" according to OECD nomenclature)—have very small overall shares. Nevertheless, each shows significant increases in share over the last decade of the 20th century. Figure C.4 shows an increase in share for many countries, with the notable exceptions of France, Japan, and Switzerland. However, globalization is more than simply an absolute increase in international activity, so it is helpful to look at patterns in transnational patent ownership. Figure

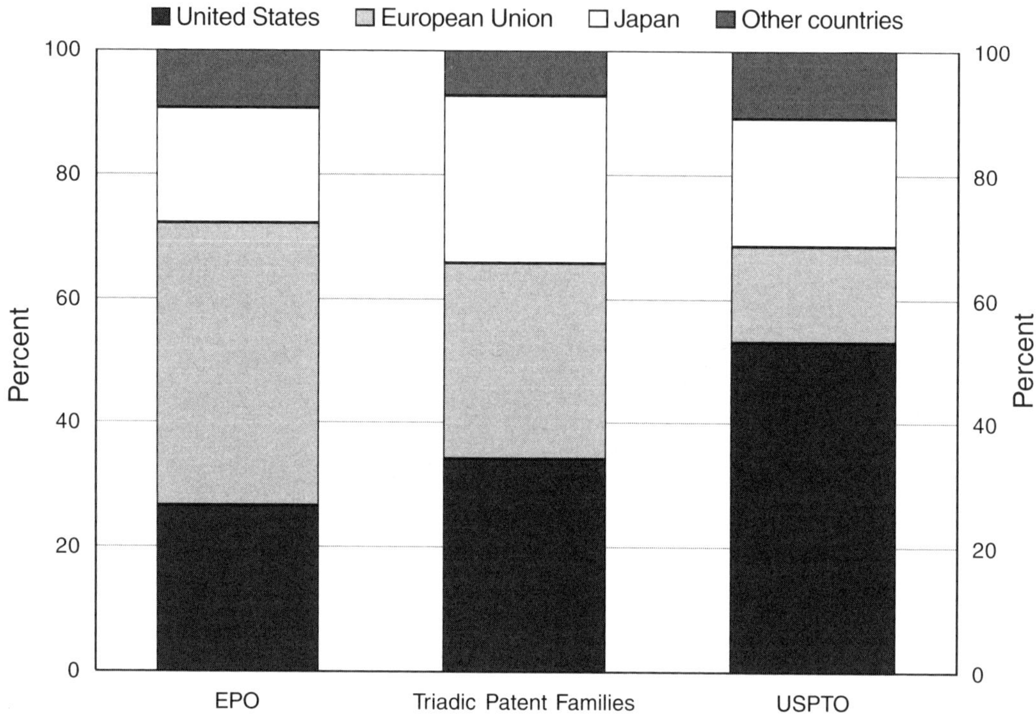

FIGURE C.3 The columns at the left and right show the geographic distribution (location of the inventor) of patent applications at the European Patent Office (EPO) and the U.S. Patent and Trademark Office (USPTO). Clearly, European inventors are the largest grouping at the EPO and U.S. inventors are the largest grouping at the USPTO. The skewing of patent data toward domestic inventors is clear. The middle column shows totals for what the OECD terms "triadic patent families," that is, inventions for which patent applications are presented at all three agencies—the EPO, the USPTO, and the Japanese Patent Office (JPO). The skewing of patent data toward domestic inventors is clear. SOURCE: OECD, *Compendium of Patent Statistics 2004*, © 2004.

C.5, Figure C.6, and Figure C.7 all indicate an increase in transnational activity—for instance, in many countries a growing share of patents is owned by firms registered in other countries, a sign of increasing global activity in R&D with concomitant increases in global and transnational ownership and collaboration.

MSE PATENTS

The data presented here are part of a limited and focused analysis of the trends in the country of residence recorded for the inventors associated with MSE patents awarded by the USPTO over the last 25 years in certain key and well-defined MSE

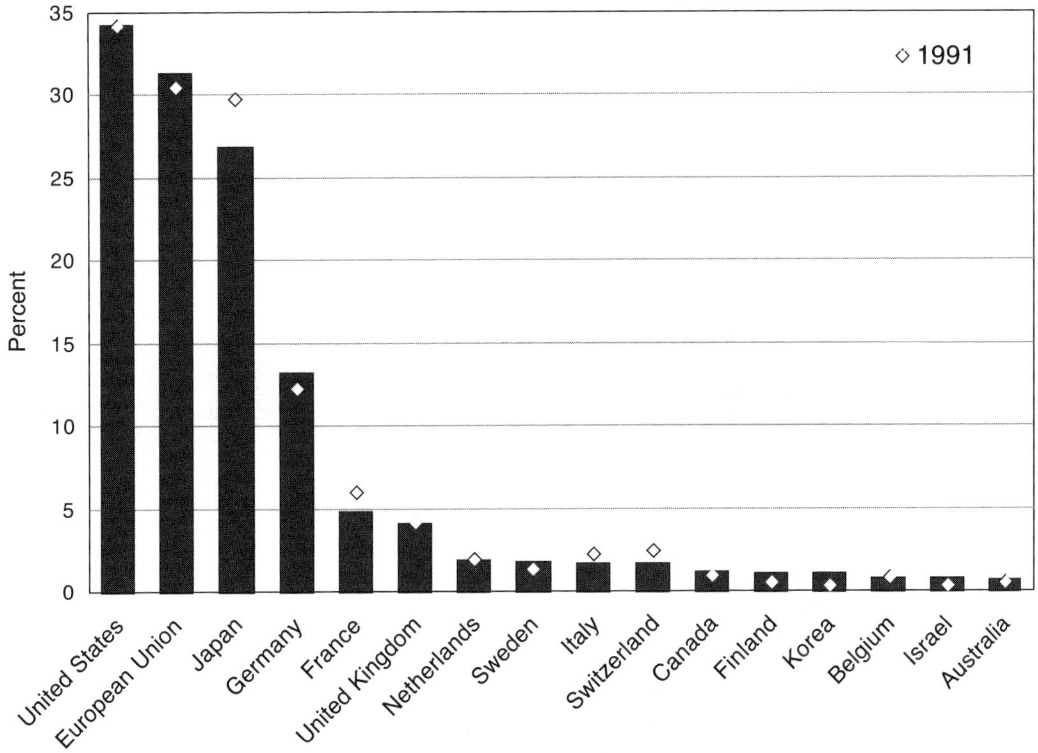

FIGURE C.4 Share of various countries in 2000 based on the country of residence of the inventors. The bars show the data for applications lodged in 2000 at the USPTO, EPO, and JPO. The diamonds show the same information for 1991. U.S. inventors are the largest grouping, accounting for 34.3 percent of applicants, similar to their 1991 share. Between 1991 and 2000 the EU's share increased by a percentage point, to 31.4 percent, whereas Japan's share declined by around 3 percentage points, to 26.9 percent. (Note that the percentages do not add to 100. This is due to the double counting of EU members.) SOURCE: OECD, *Compendium of Patent Statistics 2004*, © 2004.

subfields. Notwithstanding the weaknesses associated with a one-patent-office methodology, such a methodology can be useful in understanding the extent to which U.S. dominance has changed over time. The data were gathered from the online USPTO database.[1]

Patents of interest were identified by their associated international (World Intellectual Property Organization) category or by searching for a keyword in the patent title. The number of inventors from countries with well-established and

[1]The search engine on the EPO Web page did not allow for a similar data-mining exercise on EPO data.

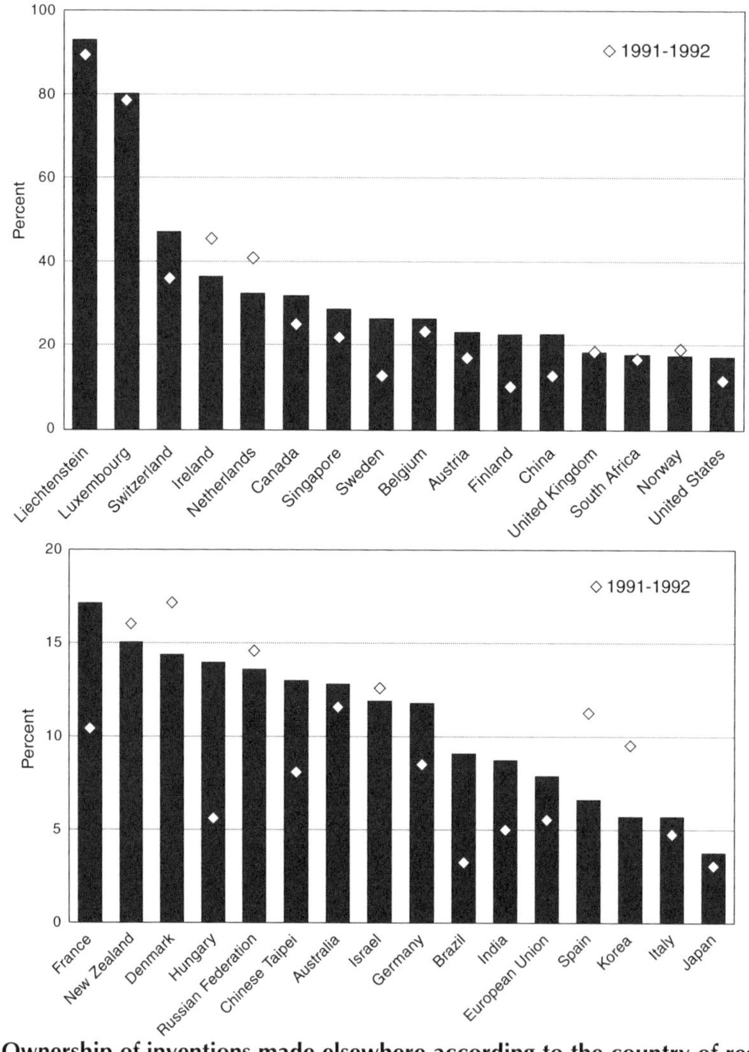

FIGURE C.5 Ownership of inventions made elsewhere according to the country of residence of the patent applicant. (The patent applicant and inventor need not be the same, and they might not be located in the same country.) The bars show the data for 1999–2000 and the diamonds show the data for 1991–1992. Note the change of scale on the left-hand axis between the upper and lower charts. These data show, for example, that over 90 percent of the patents applied for by applicants in Lichtenstein are for inventions made in other countries. The ownership of inventions made elsewhere is high in small, open economies. This is partly explained by the presence in such countries of the headquarters of multinational corporations with R&D activities elsewhere. In terms of absolute numbers, the United States and Germany are the largest owners of inventions made in other countries, but because they also generate much intellectual property domestically, their overall ranking in the above ownership data is low. SOURCE: OECD, *Compendium of Patent Statistics 2004*, © 2004.

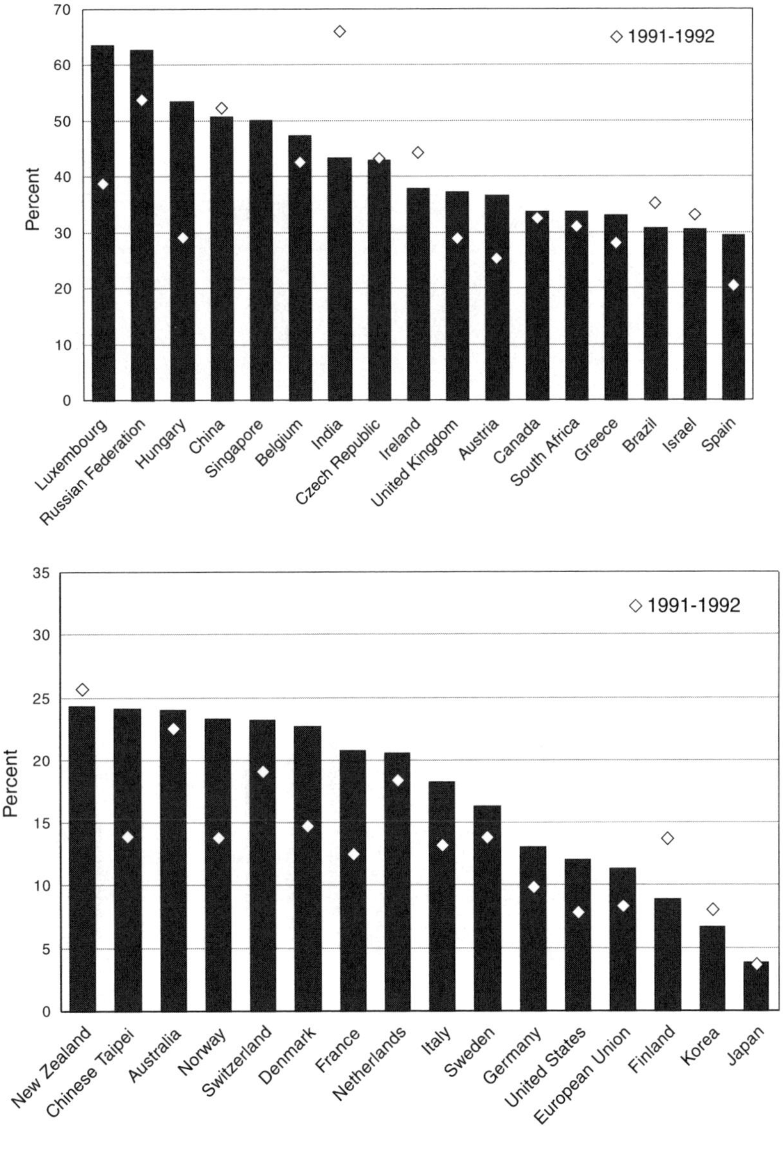

FIGURE C.6 Foreign ownership of domestic inventions according to the country of residence of the inventors. The charts are based on data from the EPO. The bars show the data for 1999–2000 and the diamonds show the same information for 1991–1992. Note the change of scale on the left-hand axis between the upper and lower charts. These data show, for example, that over 60 percent of the owners of patents for inventions made in Luxembourg are resident in another country. It is clear that as corporations relocate their R&D activities abroad an increasing share of technology is owned by entities that are not in the same country as the inventor. SOURCE: OECD, *Compendium of Patent Statistics 2004*, © 2004.

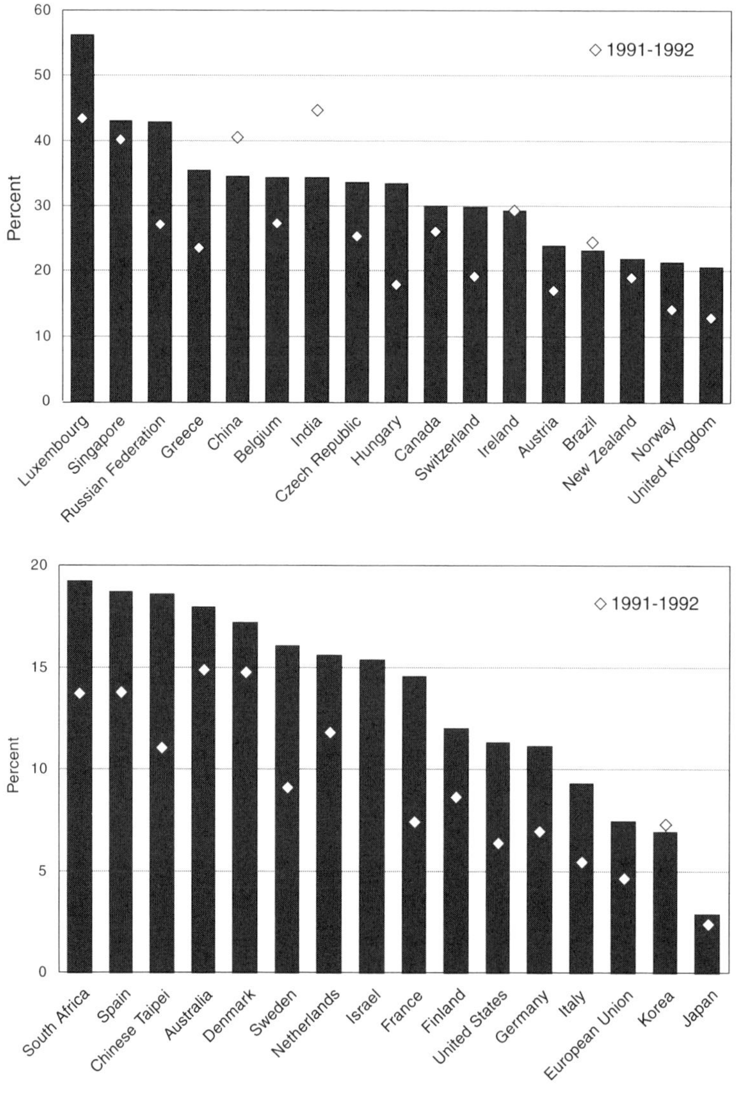

FIGURE C.7 Percentage of patents with at least one foreign coinventor, according to the residence of the inventors. The bars show the data for 1999–2000 and the diamonds show the same information for 1991–1992. Note the change of scale on the left-hand axis between the upper and lower charts. These data show, for example, that about 55 percent of patent applications from Luxembourg have at least one coinventor not from Luxembourg. NOTE: The EU is treated as one country; intra-EU cooperation is excluded. As R&D becomes a more global activity, the number of patents with a foreign coinventor should increase. The data shown here are for applications to the EPO with at least one foreign co-inventor. They show an increase in the share of patents with foreign coinventors during the 1990s. Two exceptions are China and India, where there may be an increase in domestic-only activity. SOURCE: OECD, *Compendium of Patent Statistics 2004*, © 2004.

emerging levels of activity in MSE were recorded. The subfields chosen for this analysis were alloys, catalysts, ceramics, magnetic materials, and composites. Data were gathered on the basis of the location of the inventors recorded on the patent approval record. For each subfield, the trend in the number of patents for inventors located in each country—and for the Asia5 and Euro5 compilations—is shown over a 25-year period in 5-year increments. In addition, the data are shown normalized to the U.S. number of patents. A country with a "share" of 1, measured in this way, has as many patents with inventors from that country as does the United States in that time period.

Alloys

Figure C.8 shows the distribution of patents in the alloys subfield[2] organized by the recorded country of the inventors listed in the patent database and the country-by-country share, normalized to U.S. activity in the same period.

The data indicate that patent applications in the alloys subfield are dominated by scientists and engineers in the United States, Japan, and Western Europe. The Asia5 numbers are dominated by those from Japan. The number of patents originating in the United States remained fairly steady over the last 25 years, around 550 patents a year. Japan has significantly increased its absolute number of patents (from 251 to 653 in the period reported), and its share relative to U.S. activity surged, surpassing the United States in the mid-1990s. Western Europe has had a steady increase in activity by this measure, and within the five countries surveyed, French-based inventors show the strongest increase in number of patents awarded. Western Europe's share, measured against that of the United States, has increased by 50 percent over the 25 years measured, from 30 percent to 47 percent.

It is clear from the data for all classes of patents at the USPTO over the same period (Figure C.2) that the rise in the Japanese and European share of U.S. patents in all classes vis-à-vis that of the U.S. has remained fairly steady over the last 25 years—with, if anything, a small decrease in the European share. Comparing the alloys data with the data for all classes leads to the conclusion that the Japanese and European share of activity in alloy R&D, as measured by patent data, has significantly increased in the last 25 years and that the United States has lost its leadership position to Japan. The rise in the share of Japanese- and French-originated patents is not a consequence of an overall increase in their share in all technology fields; rather, it appears that the trends in alloy patenting are different from those in patenting as a whole.

---

[2]WIPO classification C22C includes master alloys, cast-iron alloys, iron alloys, radioactive alloys, amorphous alloys, and alloys containing fibers or filaments.

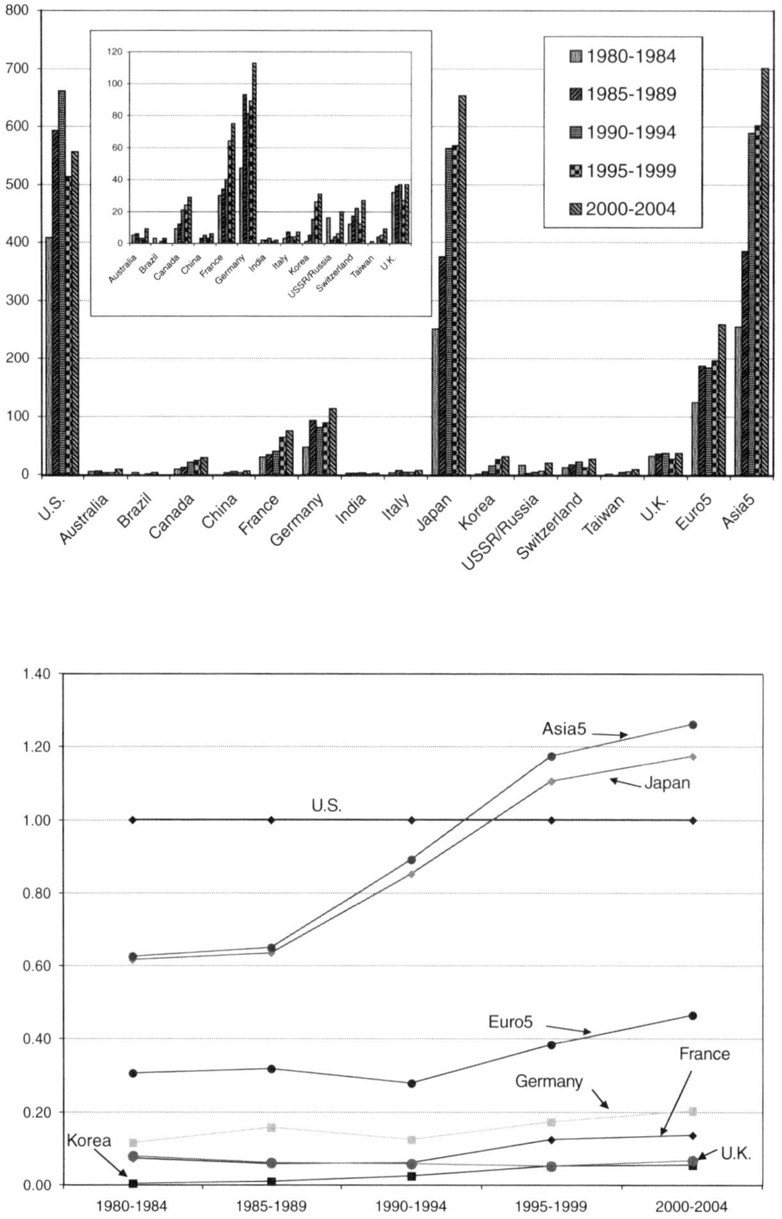

FIGURE C.8 *Above:* **number of U.S. patents awarded between 1980 and October 21, 2004, related to alloys research. Insert shows the data without Japanese, United States, and Euro5 data.** *Below:* **number of patents for each 5-year search period normalized to the number of U.S.-originated patents. SOURCE: USPTO.**

Another key question is whether there is evidence of changing activity in countries that, unlike Japan or the European countries, have not had a high profile in MSE R&D. The data for alloy research show that while the number of patents awarded with inventors in China and Korea—the two strongest nontraditional performers—has increased, the absolute numbers remain an order of magnitude behind those for Europe, Japan, and the United States. The apparent increase in activity, however, may indicate the early stages of the effects of globalization of this type of R&D in Asia—a very recent and still emerging trend. The divergence in the share factor for Japan and Asia in Figure C.8 (bottom) also indicates the emergence of activity among the four Asian countries other than Japan. It is also worth noting that the number of patents in this field from inventors from the Russian Federation surged in the last decade to levels seen in the Soviet Union in the early 1980s, which included inventors in now-independent states such as Ukraine.

Catalysts

Figure C.9 shows patent data for catalysts for the same periods and the same countries as the alloy analysis above.[3] The trends are quite different from those for alloys.

The dominance of the United States in catalyst patents is persistent and clear, with over twice as much patent activity associated with American inventors as for inventors from the European and Asian groupings. However, the number of patents awarded from Europe and Asia show significant increases relative to the U.S. numbers in absolute numbers and in share, with Japan, Germany, and Italy in particular demonstrating a surge in patent numbers and France being a strong participant in the sector but with relatively flat numbers.

The numbers of patents associated with inventors from the emerging centers of R&D—China, India, Korea, and Taiwan—all show substantial increases over 25 years ago, when they had no few or no patents in this classification. As with alloys, while the relative surge in these countries is strong, the number of patents remains an order of magnitude lower than the numbers in Japan and Western Europe and two orders of magnitude lower than in the United States. Once again Russia shows a return to the numbers it was achieving 25 years ago, but activity is slight.

[3]WIPO classification B01J includes chemical or physical processes—for example, "catalysis, colloid chemistry; their relevant apparatus (processes or apparatus for specific applications)."

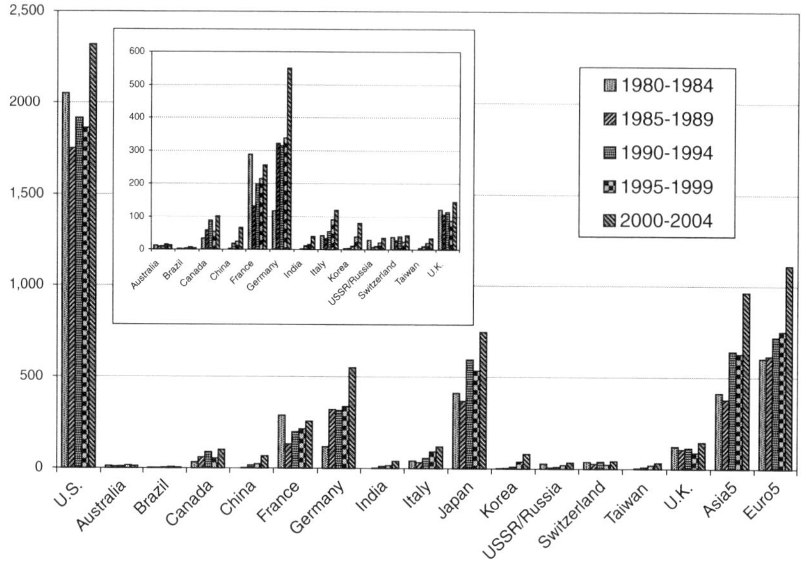

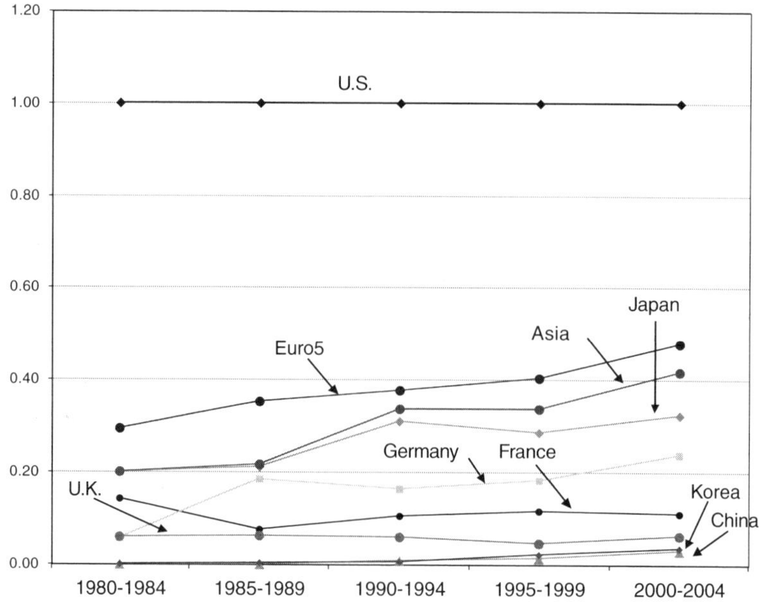

FIGURE C.9 ***Above:*** **number of U.S. patents awarded between 1980 and November 3, 2004, related to catalysis research. Insert shows the data without Japanese, U.S., and Euro5 data.** ***Below:*** **number of patents for each 5-year search period normalized to the number of U.S.-originated patents. SOURCE: USPTO.**

Ceramics

Figure C.10 shows the "ceramics" patent data at the USPTO.[4] Patenting activity is dominated in this subfield by the United States and Japan. The number of patents awarded with inventors in Japan jumped significantly at the beginning of the 1980s, with recent data indicating activity on a par with inventors in the United States. Given the home advantage from the use of U.S. patent data, it is possible that Japan has in fact equaled or surpassed the United States in the last decade. Activity in Europe is running at about half that in the United States and Japan. Germany dominates European activity but has remained fairly steady in terms of share over the last 25 years. The Euro5 grouping shows a small decrease over the same period. The share for the Asia5 grouping diverges from the Japanese share as activity has increased in the emerging four since 1990. Taiwan and Korea show the strongest increases in performance, but the absolute numbers remain very low.

Magnetics

Figure C.11, based on magnetic materials patent data from the USPTO,[5] shows significant growth in the patenting of magnetic technology over the last 20 years. Patent outputs from the United States and Japan have grown side by side, with activity over the last 5 years in both countries outperforming the combined activity of the five European countries by a factor of 4. Europe's share has decreased over the last 20 years primarily because its number of patents was static at a time when Japan and the United States were surging. The share data show a divergence between Japan and the Asia5 grouping owing to significant increases in output, primarily from Taiwan and Korea, although as with the other subfields the absolute numbers of patents with inventors from these countries remains low.

Composites

Figure C.12 shows the data for composite materials patents at the USPTO.[6] The data show an increase in global research on composite materials over the last 25 years, with the output from the United States, Asia, and Europe about on a par.

---

[4]As there was no WIPO classification with broad enough coverage, USPTO data were searched for patents with the word or part-word "ceramic" in the title.

[5]USPTO data were searched for patents with the word or part-word "magnetic" in the patent title.

[6]USPTO data were searched for patents with the words "composite" and "material" in the patent title.

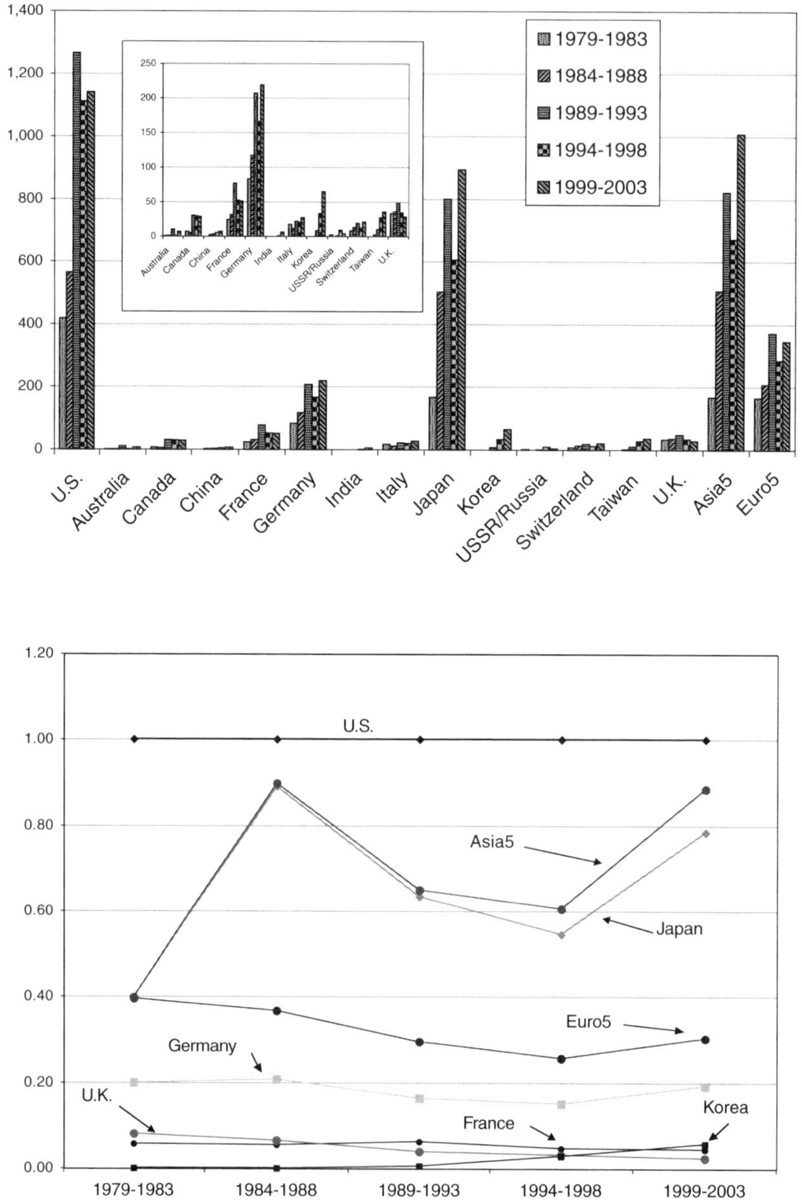

FIGURE C.10 ***Above:*** **number of U.S. patents awarded between 1980 and November 3, 2004, related to ceramics research. Insert shows the data without Japanese, U.S., and Euro5 data.** ***Below:*** **number of patents for each 5-year search period normalized to the number of U.S.-originated patents. SOURCE: USPTO.**

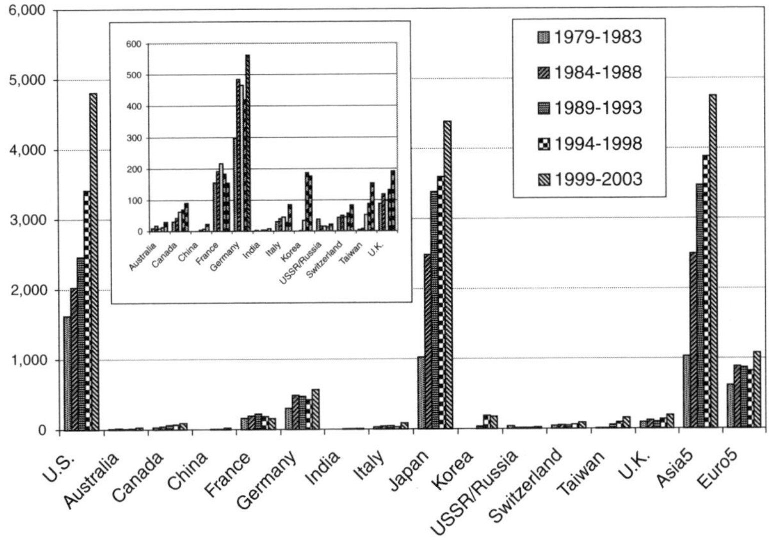

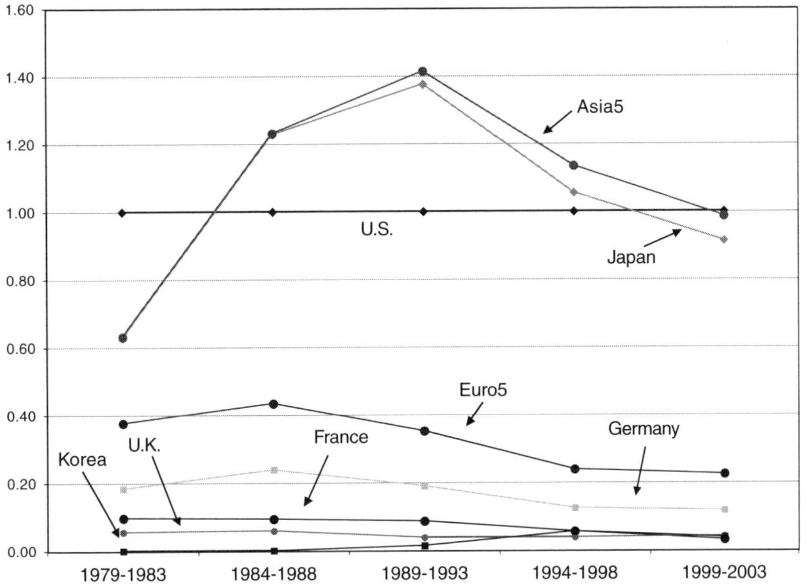

FIGURE C.11 ***Above:*** **number of U.S. patents awarded between 1979 and 2003 related to magnetic materials research.** ***Below:*** **number of patents for each 5-year search period normalized to the number of U.S.-originated patents. SOURCE: USPTO.**

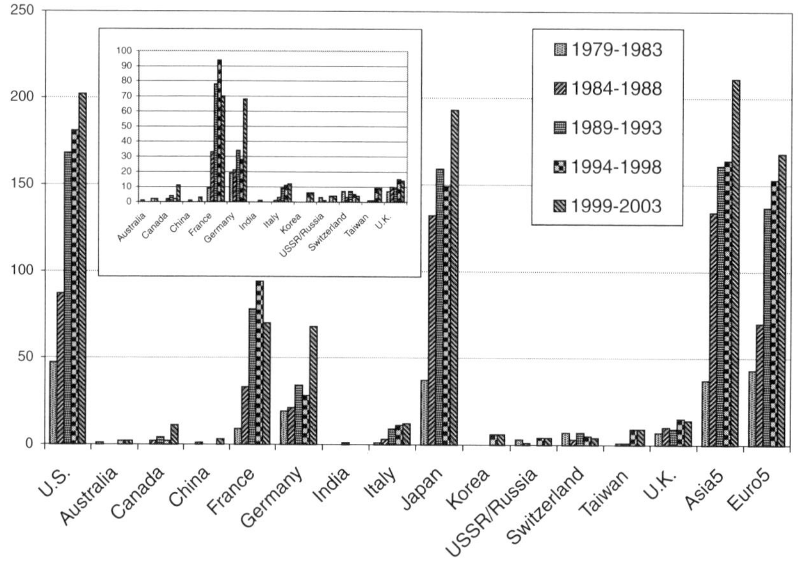

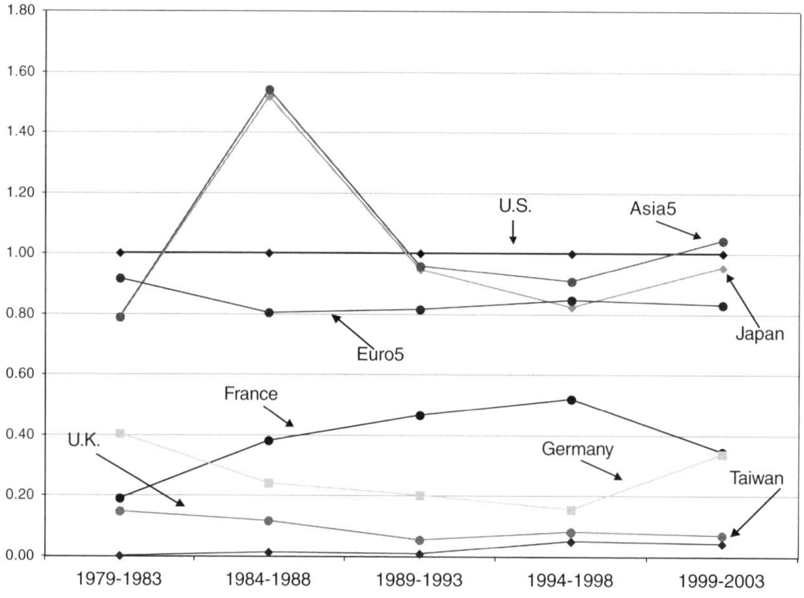

FIGURE C.12 *Above:* number of U.S. patents awarded between 1979 and 2003 related to composite materials research. *Below:* number of patents for each 5-year search period normalized to the number of U.S.-originated patents. SOURCE: USPTO.

Activity in Europe is dominated by Germany and France. Patent output with inventors from Italy shows a significant upturn, while activity in the United Kingdom and Switzerland remains static. It appears from the data that the United States lagged behind Japan in the mid-1980s but has caught up since. Again, activity in Taiwan and Korea has emerged over the last 10 years, but overall numbers remain low for both countries. Compared with other subfields, research in composites is occurring in only a small number of countries, with emerging actors such as China and India having displayed no significant U.S. patenting activity.

GLOBALIZATION TRENDS AS SEEN FROM PATENT DATA

Although there are significant differences in the indicators for the five subfields chosen for this analysis, some general trends emerge:

- In each subfield the United States is the world leader or among the world leaders.
- Japan and Western Europe are the closest rivals for leadership. Japan appears to have surpassed the United States in alloys and appears ready to surpass the United States in ceramics.
- Global activity in all the subfields examined is diversifying, with significant increases in activity in Asian countries that so far have not had substantial activity in these fields.
- Notwithstanding the global diversification of research activity, the number of patents associated with inventors from the emerging centers remains low.

D

Global Trends in Literature Authorship

Data were gathered from the Science Citation Index Expanded format.[1] The article-title field of the database was searched for relevant words in each of the areas chosen and the number of articles found from each country recorded.

ALLOYS

Figures D.1 and D.2 show the literature data for the alloys subfield from 1984 to 2003. The global output of scientific papers has grown considerably in those 20 years, but while the number of papers from the United States has increased steadily, its share of the global numbers has decreased. The same is true for Japan, whereas the share of global output for the Euro5 and the Asia5 groups has steadily increased, as has their performance when matched against the United States—as shown in the top chart of Figure D.2. In particular, the output of papers with authors from China and Korea shows exceptional growth, with recent numbers from China approaching those of the United States and Japan and outstripping more traditional centers of R&D in countries like Germany, France, and Britain.

[1]The Science Citation Index Expanded provides access to current and retrospective bibliographic information, author abstracts, and cited references in approximately 5,900 of the world's leading scholarly science and technical journals covering more than 150 disciplines. For further information, see http://www.isinet.com/products/citation/scie/.

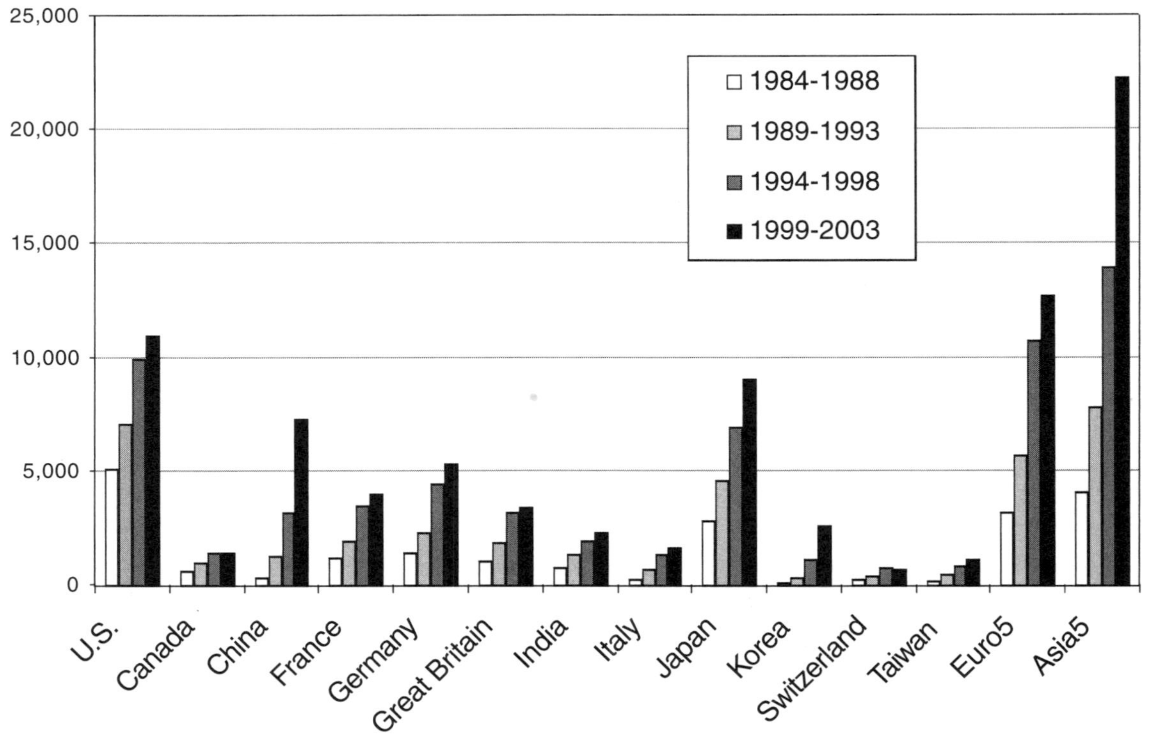

FIGURE D.1 Country of authorship for 20 years of literature (1984–2003) in the alloys subfield. Chart shows the total number of papers for each country and for the Asia5 and Euro5 groups.

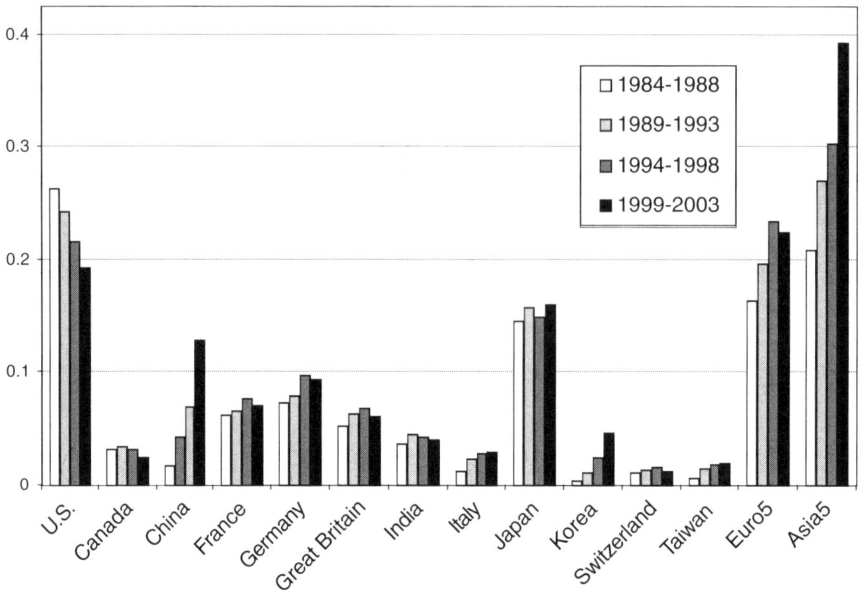

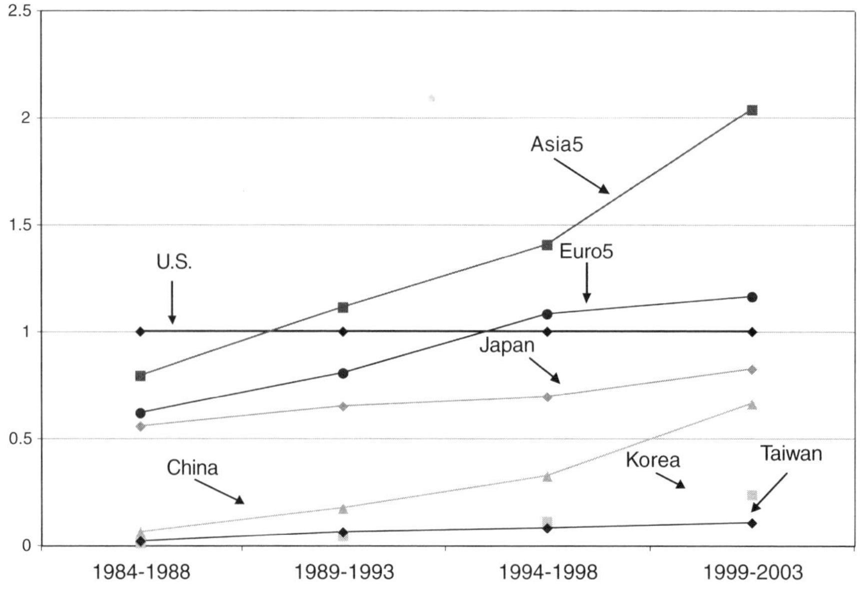

FIGURE D.2 Country of authorship for 20 years of literature (1984–2003) in the alloys subfield. The upper chart shows the share for each country as a fraction of the total number of papers for that time period. The lower chart shows the share factor normalized to the literature output of authors in the United States.

CATALYSIS

Figures D.3 and D.4 show general growth in the global output of literature in the subfield of catalysis over the 20 years. However, the data also show that despite the strong increase in the number of papers authored in the United States, the U.S. share of the global total declined somewhat after 1990. The United States does not enjoy any discernible and reliable lead in the catalysis field, although it is among the leaders. Japan's share has remained steady, while the numbers from the Euro5 countries showed a small but discernible increase in the 1990s and some stagnation more recently. The Asian share of the literature output, however, shows a steady and strong increase over the 20 years, outstripping U.S. output from 1999 to 2003. China shows a particularly strong increase, surpassing the individual European countries and approaching Japan's output. The increase has been significant for Korea but less so for Taiwan.

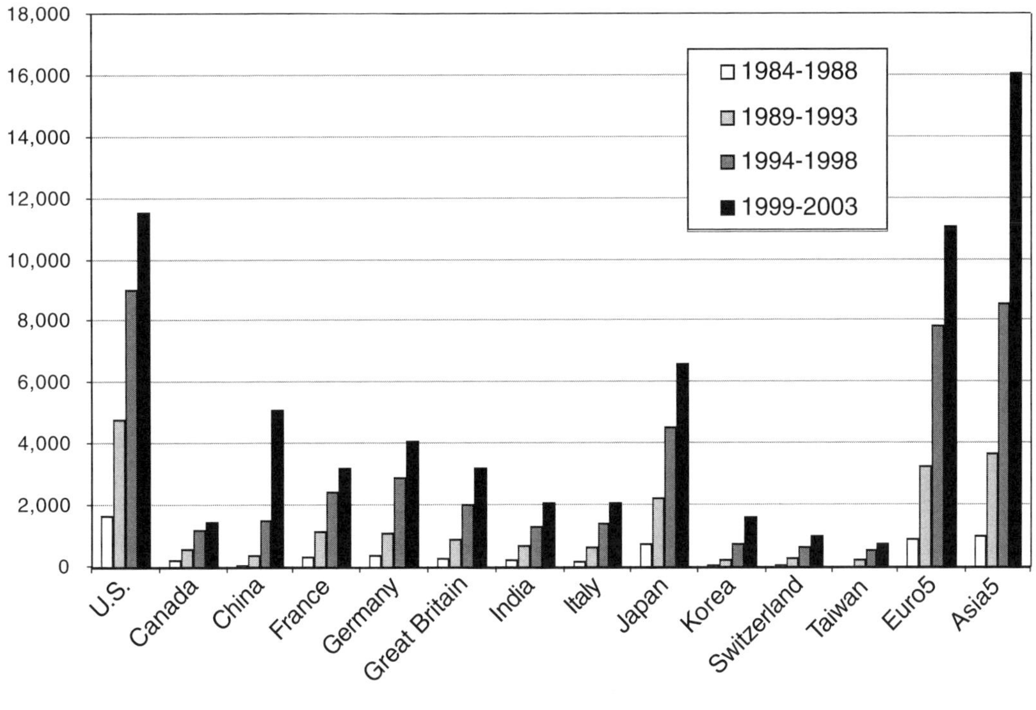

FIGURE D.3 Country of authorship for 20 years of literature (1984–2003) in the catalysis subfield. The chart shows the total number of papers for each country and for the Asia5 and Euro5 groups.

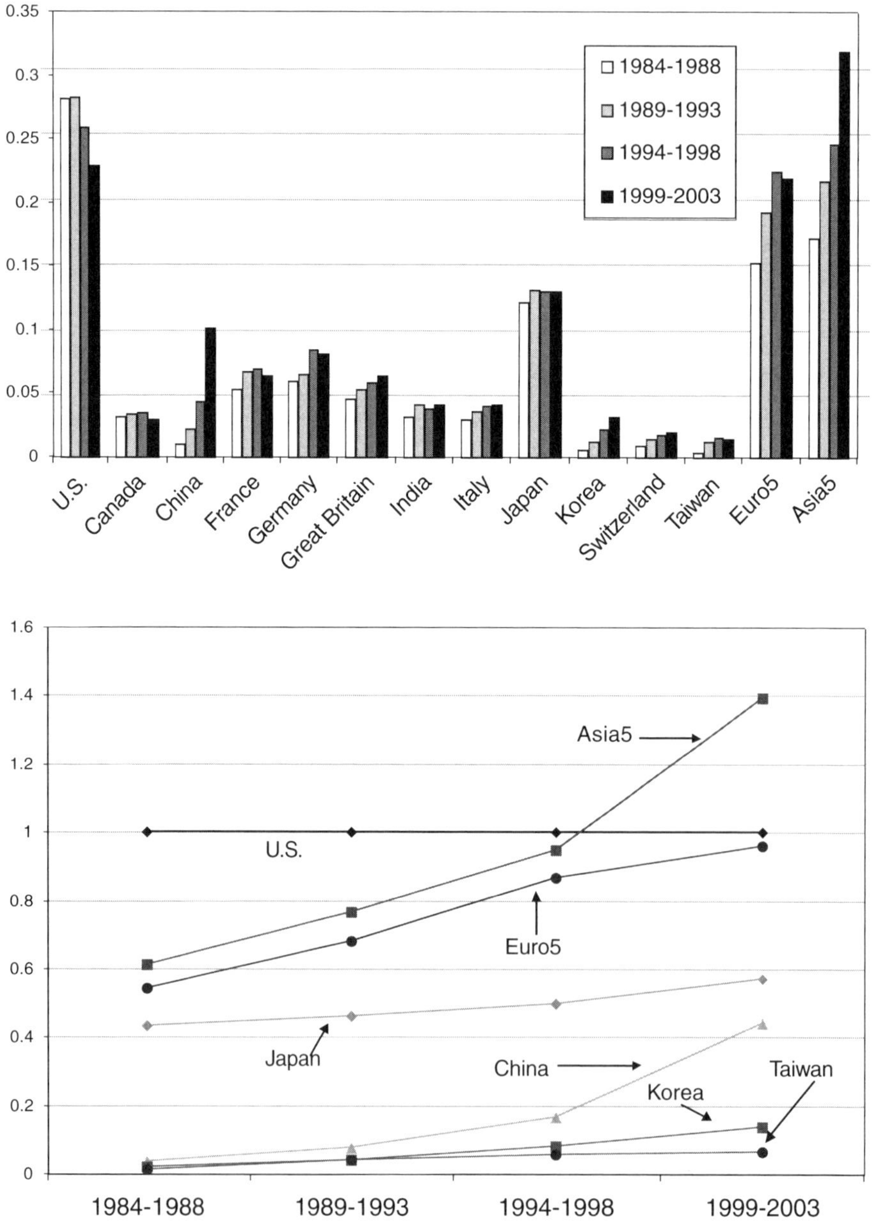

FIGURE D.4 Country of authorship for 20 years of literature (1984–2003) in the catalysis subfield. The upper chart shows the share for each country as a fraction of the total number of papers for that time period. The lower chart shows the share factor normalized to the literature output of authors in the United States.

COMPOSITE MATERIALS

The literature on composite materials flourished from 1984 to 2003 across the globe (Figures D.5 and D.6). The United States remains in the lead in terms of share of global total, but this leadership was challenged between 1999 and 2003 by the European and Asian regions. In Europe, Britain enjoys the lead, with Germany surging ahead to nearly equal the output from France or Britain. China's share is approaching Japan's, and significant increases can be seen in the output share for Taiwan and Korea, although the numbers are small.

OPTICAL-PHOTONIC MATERIALS

Data in Figures D.7 and D.8 show the United States maintaining its lead in literature over the 20-year period, but with the European and Asian regions approaching it over the last 15. Italy shows the strongest surge in Europe of the countries

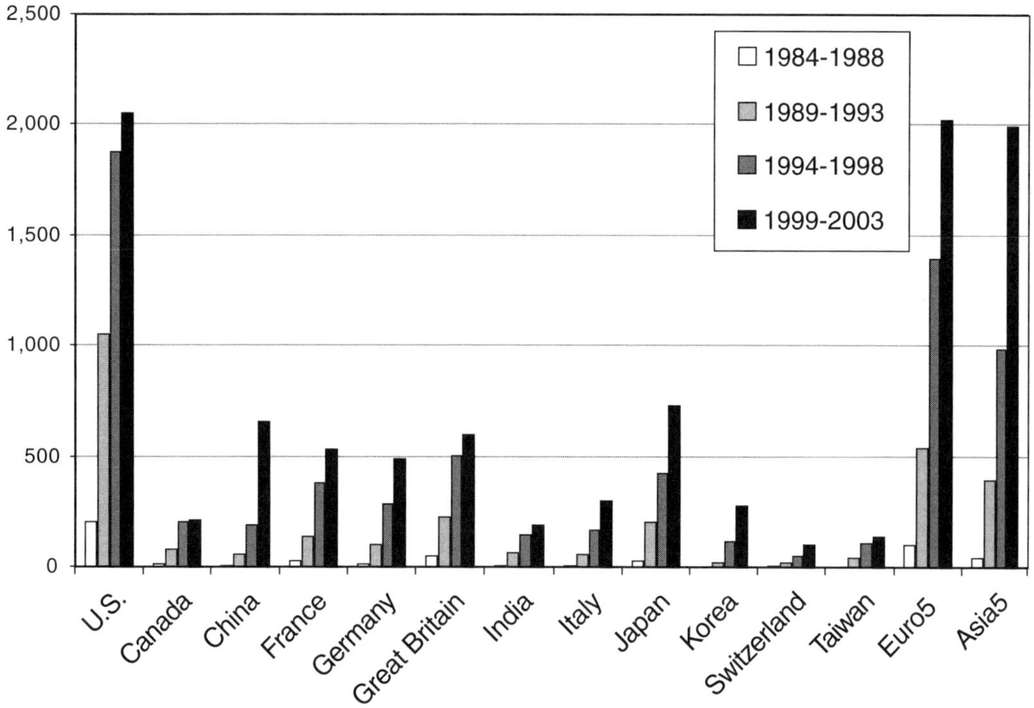

FIGURE D.5 Country of authorship for 20 years of literature (1984–2003) in the composites subfield. Chart shows the total number of papers for each country and for the Asia5 and Euro5 groups.

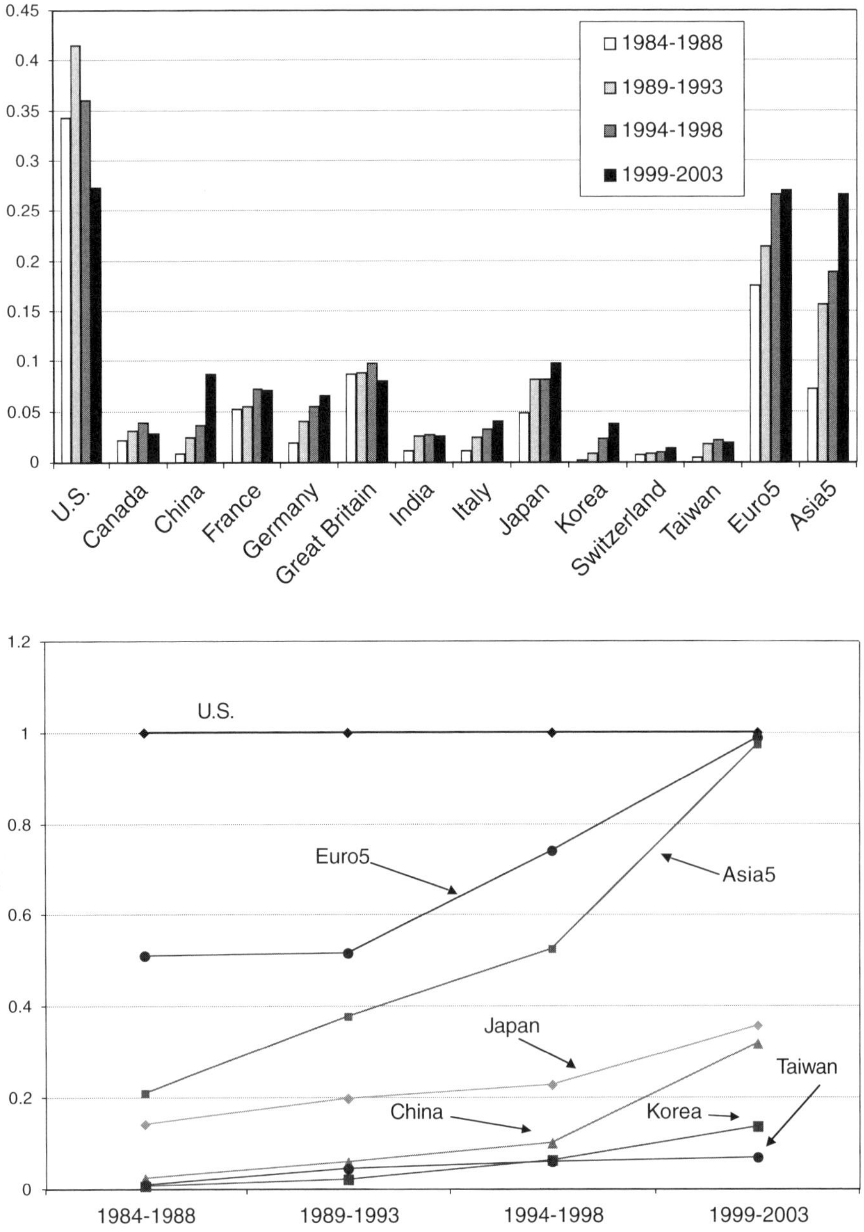

FIGURE D.6 Country of authorship for 20 years of literature (1984–2003) in the composites subfield. The upper chart shows the share for each country as a fraction of the total number of papers for that time period. The lower chart shows the share factor normalized to the literature output of authors in the United States.

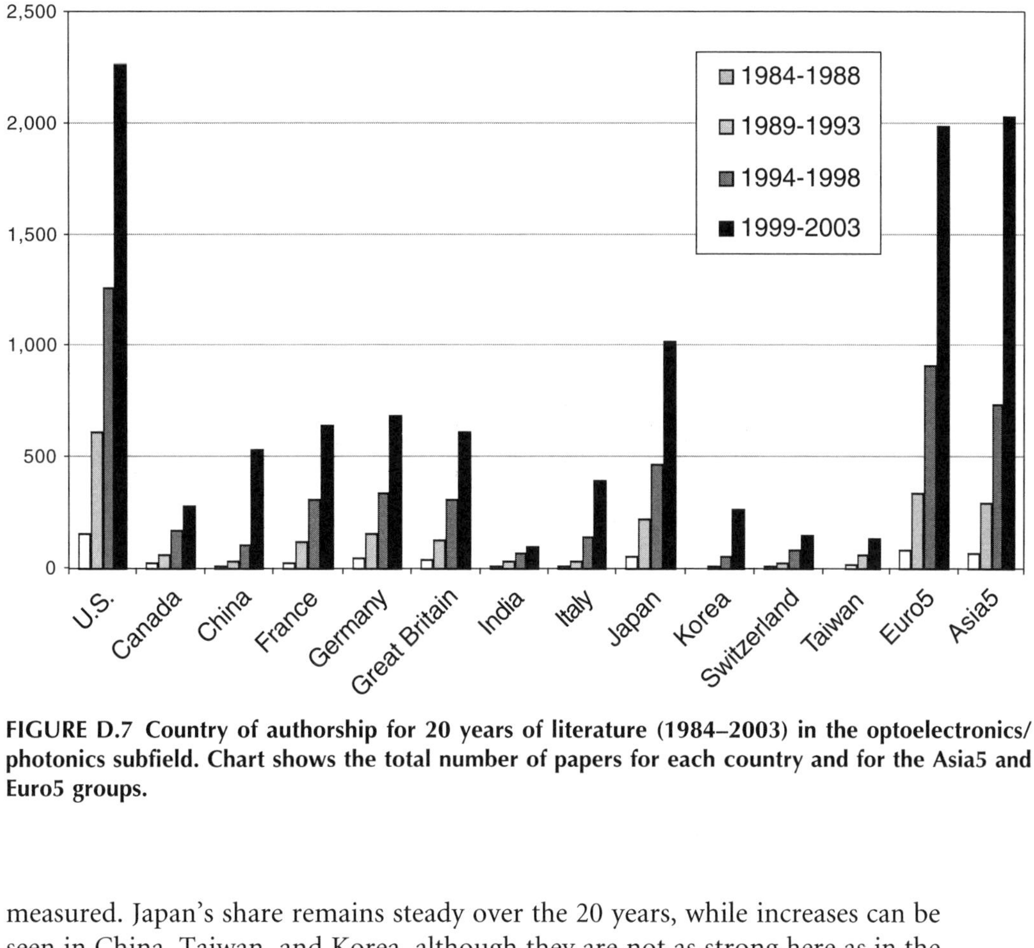

FIGURE D.7 Country of authorship for 20 years of literature (1984–2003) in the optoelectronics/photonics subfield. Chart shows the total number of papers for each country and for the Asia5 and Euro5 groups.

measured. Japan's share remains steady over the 20 years, while increases can be seen in China, Taiwan, and Korea, although they are not as strong here as in the other subfields.

GENERAL TRENDS IN GLOBALIZATION FROM LITERATURE DATA

Although there are differences between the various sets of data, some general trends emerge:

- The United States is among the world leaders but not *the* clear leader in all four subfields.
- Western Europe and Japan are also among the leaders in these subfields and showed clear surges in activity in the subfields from 1989 to 2003, overcoming previous U.S. dominance.

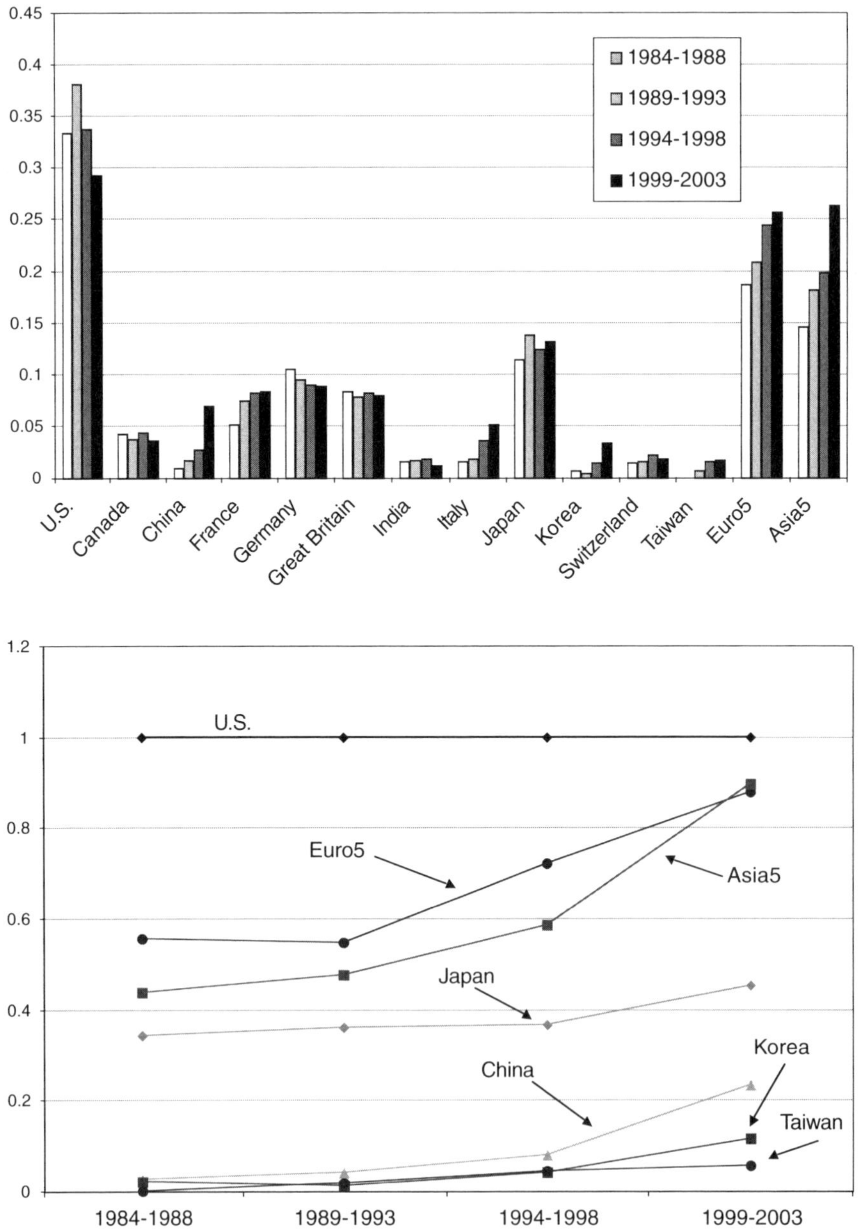

FIGURE D.8 Country of authorship for 20 years of literature (1984–2003) in the optoelectronics/photonics subfield. The upper chart shows the share for each country as a fraction of the total number of papers for that time period. The lower chart shows the share factor normalized to the literature output of authors in the United States.

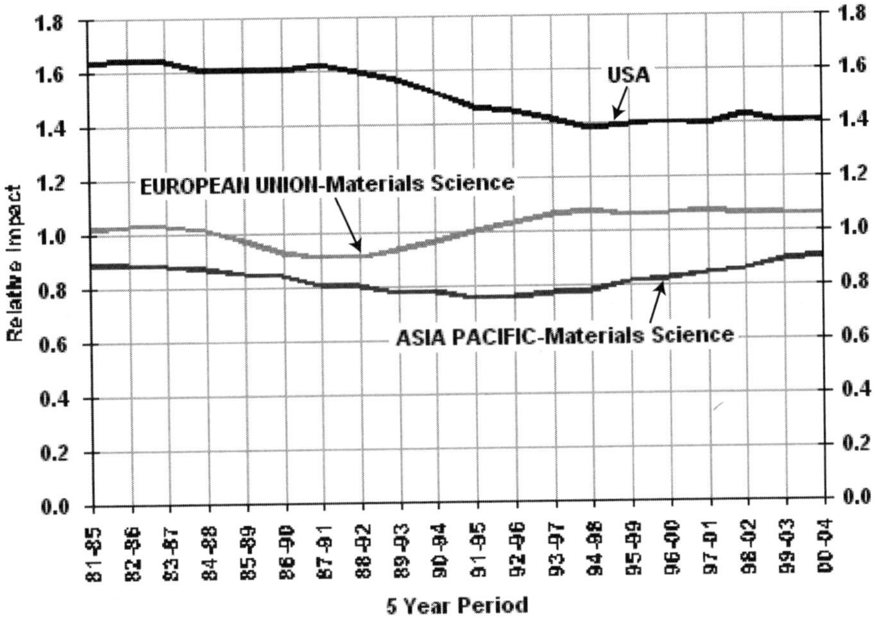

FIGURE D.9 Relative impact of materials science papers from the United States, the Asia-Pacific region, and the European Union. SOURCE: *Thomson Scientific National Science Indicators*, 1981–2004.

- In all four subfields there has been a significant increase in the literature presence of the Asian countries included in these searches: Most notably, the number of papers from China increased substantially over the last 5 to 10 years, to levels at times approaching the number from Japan and several Western European countries.
- The globalization of authorship in the world's journals appears to be advancing rapidly, with a clear shift toward authors from countries that were not traditionally involved in these subfields.

With literature analyses, there is always a question about the quality of the papers being counted. Figure D.9 tries to ascertain the relative quality of the papers in materials science from 1981 to 2001 by counting the number of times a paper was cited. Like the preceding information, the figure shows that the U.S. lead in terms of citations has narrowed since 1980, with the Asia/Pacific region and Europe encroaching on traditional U.S. dominance.

E

Results of the Community Poll

To understand better some of the globalization trends as perceived by members of the materials community, a Web-based poll was organized and completed over a 2-week period, November 8–23, 2004. With the assistance of materials professional societies, an e-mail announcing the poll was circulated to their members, who came from institutions in the United States and overseas (Figure E.1).

It is important to recall when considering the results of the poll that the exercise is based on a self-selected group of 719 respondents. While the range of societies assisting in this exercise was broad (the American Ceramic Society, the American Physical Society, the Federation of Materials Societies, the Materials Research Society, the Society for Biomaterials, the Society of Manufacturing Engineers, and The Minerals, Metals and Materials Society), it cannot be claimed with certainty that the self-selected group is representative of the MSE community, which is increasingly interdisciplinary. The poll should, therefore, be considered to reflect in a general way the global nature of MSE activity. A more reliable poll would require the application of statistical criteria to achieve a response group that is statistically relevant. Figures E.2 through E.7 analyze the results of the questionnaire.

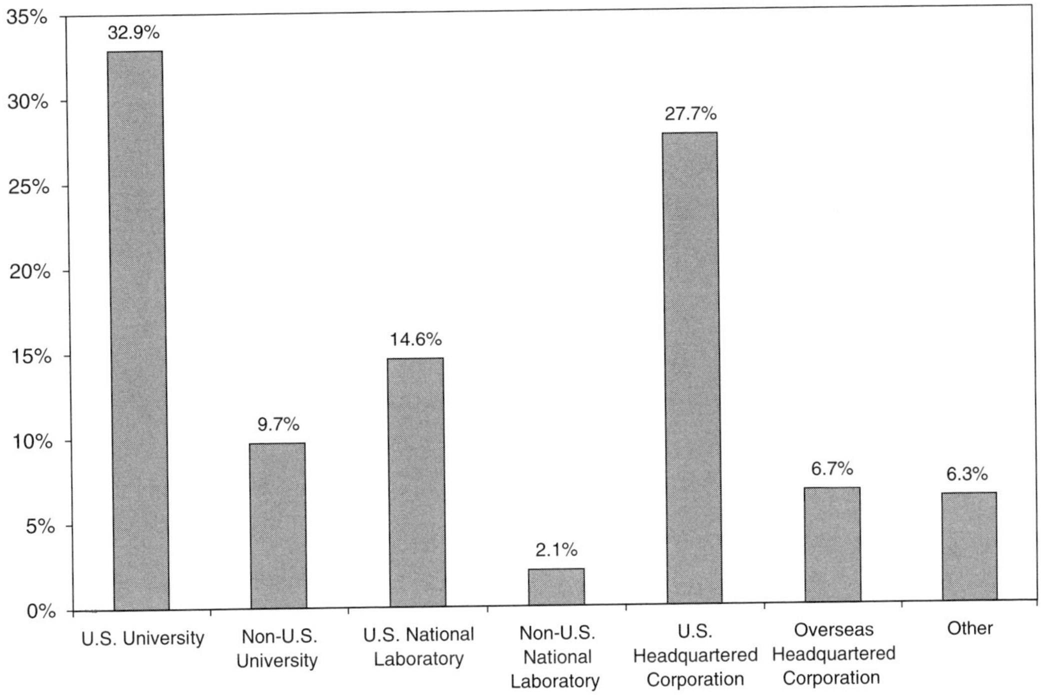

FIGURE E.1 Affiliation of respondents. The category "other" includes respondents who gave consultancy (6 respondents), nonprofit organization (5), or other U.S. government (4) as their affiliation. (Note: these data are indicative only and not based on a statistically relevant sampling.)

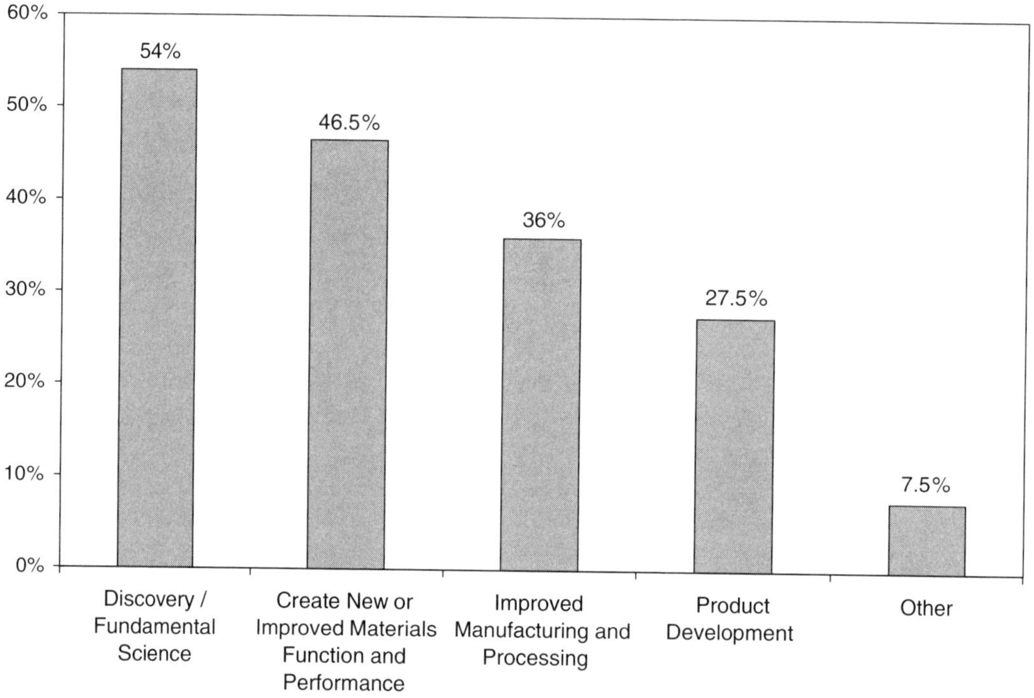

FIGURE E.2 Primary purpose of the R&D conducted at the institutions of the respondents. "Other" includes education (10 respondents) and consulting (5). (Note: these data are indicative only and not based on a statistically relevant sampling.)

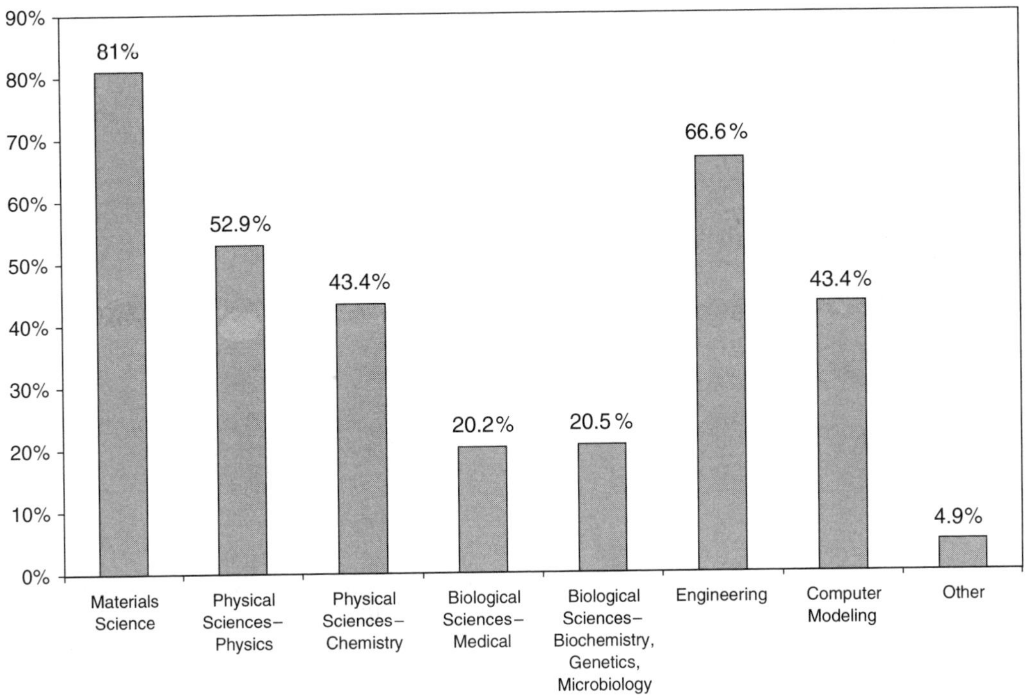

FIGURE E.3 Disciplines of interest at the institutions of the respondents. More than one answer was allowed. "Other" includes biomaterials of some kind (9 respondents) and metallurgy (5). (Note: these data are indicative only and not based on a statistically relevant sampling.)

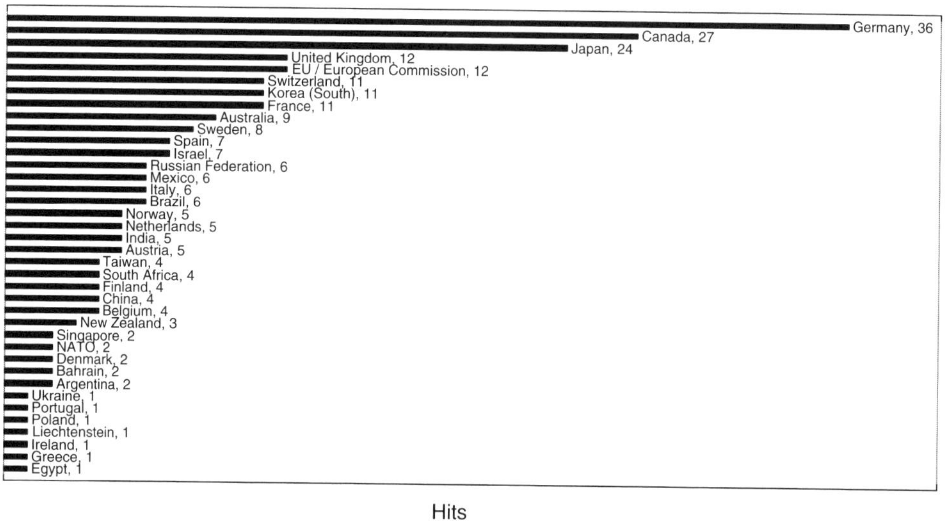

FIGURE E.4 Countries identified by both U.S. and non-U.S. respondents as non-U.S. supporters of their MSE R&D. This information can be interpreted as an indicator of the world players in R&D funding. (Note: these data are indicative only and not based on a statistically relevant sampling.)

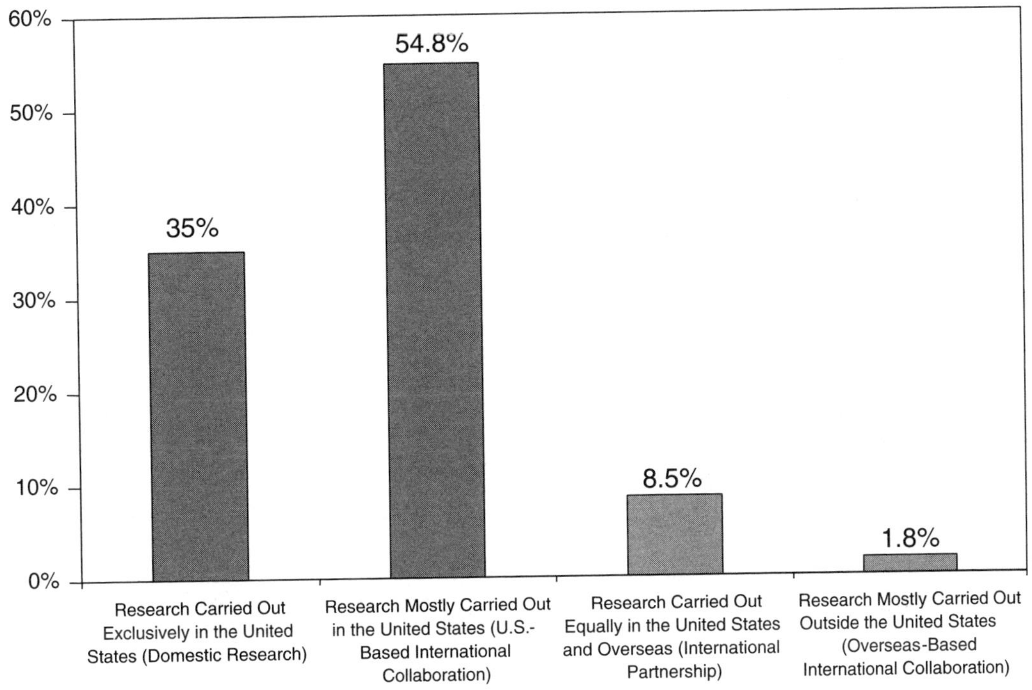

FIGURE E.5 Nature of the collaboration. Thirty-five percent of U.S.-based respondents say their research is carried out exclusively in the United States. Most U.S.-based researchers (65 percent) are involved with international collaborations, with the majority of work being carried out in the United States and in academic-academic collaborations. There is significant internationalization of U.S. corporate research, with nearly 36 percent of the research being U.S. corporate research with a foreign partner of one type or another. (Note: these data are indicative only and not based on a statistically relevant sampling.)

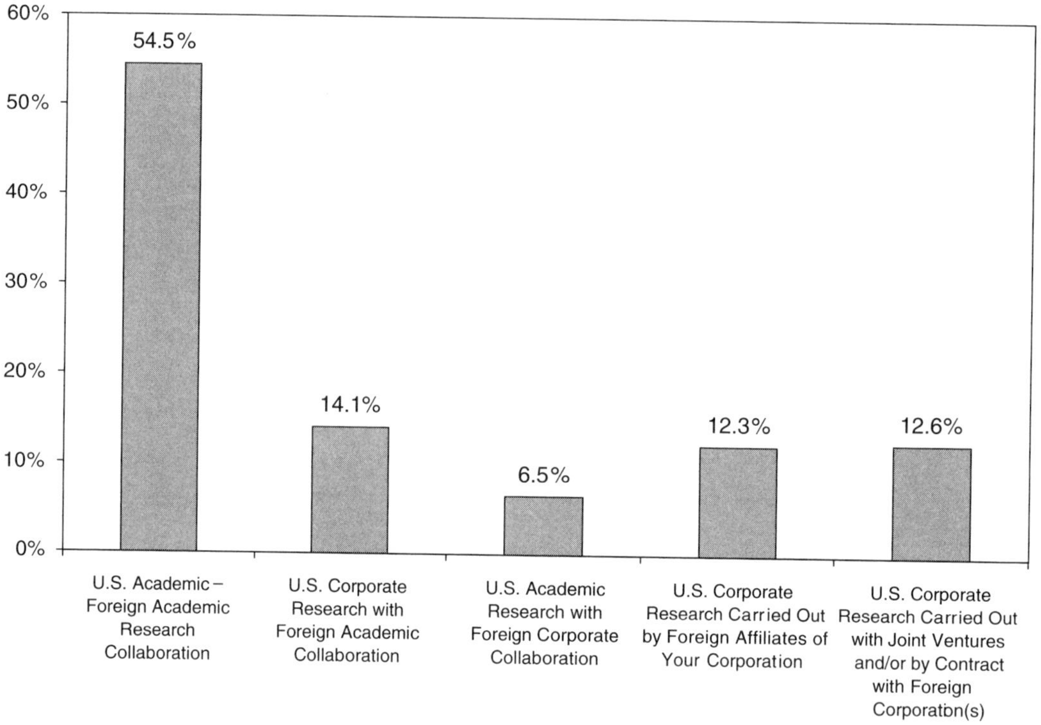

FIGURE E.6 Nature of the international collaboration as described by U.S. respondents. Most of the reported activity took place between U.S. and non-U.S. academic institutions. (Note: these data are indicative only and not based on a statistically relevant sampling.)

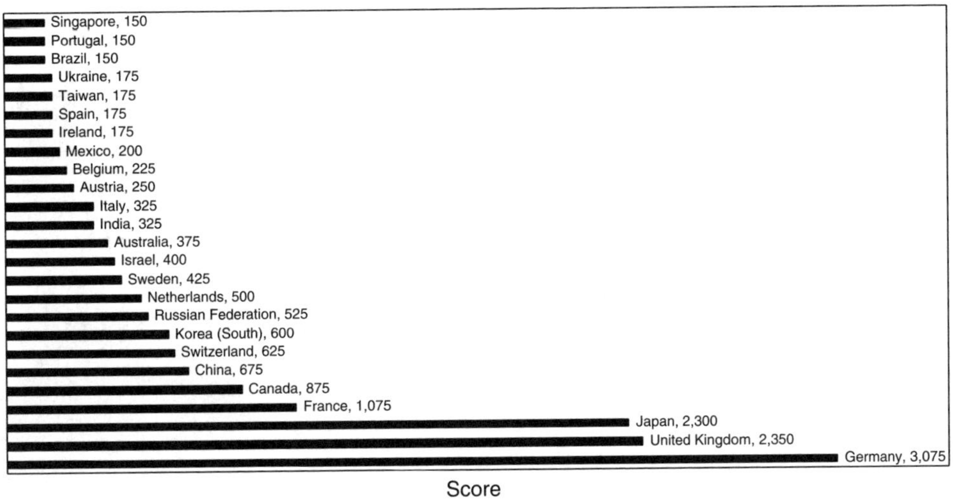

FIGURE E.7 Top three international partners. The committee weighted the first choice by an arbitrary score factor of 100, the second choice by 50, and the third choice by 25. The rankings above are based on the total weighted scores. International partnerships dominated by activity with Germany, Japan, and the United Kingdom. (Note: the rankings are indicative only and not based on a statistically relevant sampling.)

F

Superalloy Case Study

BACKGROUND

This case study examines the status of the superalloy industry in the United States and abroad today. It is based on data gathered in personal interviews[1] and from other sources.[2] In particular, this case study examines companies in the United States that melt and produce superalloy materials, that cast parts from superalloys, or that use superalloys. It also identifies other entities conducting superalloy R&D in the United States and superalloy R&D operations outside the United States. For the purpose of this study, superalloys are defined as alloys based primarily on

[1]Private discussions were held with the following individuals between March and October 2004: Rodney Boyer, The Boeing Company; Peter Bridenbaugh, Alcoa and Precision Castparts; John DeBarbadillo, Special Metals; Anthony Giamei, Pratt & Whitney; Richard Kennedy, Allvac; Dwayne Klarstrom, Haynes Stellite; Gernant Maurer, Cartech and Special Metals; Donald Muzyka, Special Metals; Fred Pettit, University of Pittsburgh; David Poirier, University of Arizona and SMPC; John Radovich, Purdue University; John Tunderman, INCO and Haynes Stellite; James Williams, GE and Ohio State University; and Frank Zanner, Sandia National Laboratories and the Special Materials Processing Consortium (SMPC).

[2]SEC filings by companies; a white paper by the Specialty Steel Industry of North America, submitted as congressional testimony in September 2004; and proceedings of the Seven Springs conferences, sponsored by The Minerals, Metals and Materials Society, 1978–2004.

nickel and cobalt that have very useful properties at elevated temperatures and/or in corrosive environments.

The superalloy industry serves a limited market, albeit with critically important products. Superalloys find application in the aerospace industry as the primary enabling material in the hot end of jet engines with both rotating and static components and in auxiliary power units for aircraft. They are also used in land-based industrial gas turbines, in petrochemical refining facilities where elevated temperatures prevail, in chemical plants where corrosive conditions exist for which normal stainless steels are not useful, and in sour oil and gas wells.

The principal cost element in the production of superalloys is the cost of the nickel or cobalt. Since none of the worldwide producers of superalloy products produces nickel, all have to buy their raw material on the open market and from the same primary or secondary (scrap) suppliers. Maintaining a competitive position therefore relies on managing operating costs and maintaining a competitive lead in state-of-the-art technology, both of which require continuing R&D on new and constantly evolving compositions and processing technologies.

The three industries that consume the most superalloy products are the aerospace industry—principally jet engines—the land-based turbine business, and the chemical process industry. Each is facing challenges that could threaten superalloy R&D in the United States. Increased international competition and globalization might threaten the aerospace industry, which has been the source of high-value exports for the United States. The land-based turbine business has grown significantly in the past decade but may be self-limiting because the rising cost of the natural gas used to generate electricity—for instance, the price of gas tripled in 2004—decreases the incentive to add natural-gas-based electricity generation to the power grid. The chemical process industry has not been investing in the United States in recent years, and if that trend continues, the market for the superalloy products that it consumes in plant construction and maintenance will shrink, putting further downward pressure on costs and profits.

COMPANIES MELTING AND PRODUCING SUPERALLOY MATERIALS IN THE UNITED STATES

Special Metals

Based principally in New Hartford, New York, Special Metals Corporation (SMC) makes forged bar products—particularly for rotating parts of jet engines—and flat-rolled nickel-base alloys. SMC became a public company in 1961 and then became private once again in 1987, when it was purchased by a French company, Albert and Duval. In 1997, SMC went public once again. In 2002, after two successive expansions—in 1998, it purchased Wiggin Alloys in the United Kingdom and INCO Alloys International[3]—it filed for Chapter 11 bankruptcy protection. SMC recently emerged from Chapter 11 and is owned by a group of banks. Before and during the bankruptcy, SMC cut spending and reduced the R&D workforce, from 50–60 persons in the mid-1990s to about 5 engineers in 2004.

Allvac

A subsidiary of Allegheny Technologies, Allvac is based in Monroe, North Carolina. A company that has had some financial difficulties, Allvac is primarily a producer of forged bar products for the rotating parts of jet engines and land-based turbines. The company's U.K. operation, Allvac Ltd. (formerly Sheffield Forgemasters) is a production facility and does no R&D.

Between 1990 and 1998 Allvac increased R&D staff from 20 to 30. Since 1998, it has cut staff to 14, only 7 of them engineers. Recently, Allvac shifted its focus from customer technical service and process improvements to new product development work. This alloy research has been funded by the USAF Metals Supportability Initiative, which led to the improved alloy 718+. The National Institute of Standards and Technology (NIST) and the Defense Advanced Research Projects Agency (DARPA) are supporting research into new processing technology.

Carpenter Technology (Cartech)

With perhaps a broader-based product line than either SMC or Allvac, Cartech —based in Reading, Pennsylvania—has continued to improve the melting and casting processes. It produces superalloy bar products and is financially stable, but it is primarily a producer of stainless steels and tool steels. Cartech has cut back on R&D staff, from about 40 people in the 1990s to 20 today, 10 of whom are engineers but do virtually no new alloy development.

Haynes International

Originally called Haynes Stellite, this Kokomo, Indiana, company was first owned by Union Carbide Corporation, then by Cabot Corporation, before becoming privately held by the Blackstone Group in 1989. The company is highly leveraged, with over $100 million in negative equity and a very large debt in the

[3]Until 1983, when International Nickel Company (INCO) closed its Sterling Forest, New York, laboratory, it was the primary developer of nickel-base superalloy compositions. When it closed, INCO employed about 350 people at its laboratory.

form of publicly held debentures. It is a producer of nickel- and cobalt-based alloys in the form of plate and sheet at its Kokomo plant and of tubular products at its Mississippi plant. Its principal competitors in the United States are Allegheny Technologies and SMC. Outside the United States, its largest competitor is Krupp VDM, in Germany. Haynes has a larger customer base than SMC or Allvac since it primarily produces corrosion-resistant alloys for the petrochemical and basic chemicals industries. However, the petrochemical industry has not built a new refinery in the United States for over 25 years, limiting the market somewhat to replacement products.

From 1980 to 2004, Haynes's R&D workforce decreased from 125 to 32, 12 of them engineers. R&D spending was reduced from $3.7 million in 2002 to an annualized rate of $2.4 million in 2004. Haynes continues to develop cobalt-based C288 alloy for jet engines, as well as new corrosion-resistant alloys such as G35 and C22HS, and it holds patents for five alloys of commercial value.

Allegheny Technology

A flat-rolled producer of primarily stainless steels, Allegheny Technology produces some superalloys, such as C276, at its plant in Brackenridge, Pennsylvania. It has cut back on its R&D activity in the past several years.

COMPANIES CASTING PARTS FROM SUPERALLOYS

Howmet

A wholly owned subsidiary of Alcoa, Howmet continues to produce investment-cast turbine blades and continues to do process research.

Cannon-Muskegon

Cannon-Muskegon became a wholly owned subsidiary of SPS Technology in 2001, which in turn was purchased by Precision Castparts in 2004. It obtains funding for its R&D from the French company Snecma and is involved in R&D with GE through a GE-Snecma partnership in engine development.[4]

[4]GE works with Snecma (France) in a joint venture called CFM International. In 2005 the Safran Group was formed following the merger of Snecma with Sagem. Sagem, a high-technology group, is the second largest French telecommunication company and the third largest European electronics company. The Snecma-Sagem merger resulted in the privatization of Snecma.

Precision Castparts

Based in Portland, Oregon, Precision Castparts now owns Cannon-Muskegon, Wyman-Gordon, and Western Australia Special Alloys (WASA), making it a dominant producer of superalloy finished parts using both the investment casting and forging routes. Precision Castparts does not do any alloy development research but does work to improve productivity and processes. It supplies some of its materials from WASA (see below). It is also investing in investment casting facilities in China.

COMPANIES USING SUPERALLOYS

GE

Continuing to conduct and support R&D, GE Engines has taken over the R&D that was done at the GE corporate research lab in the past. GE has a joint venture with Snecma (France), CFM International, and suppliers of materials such as Cannon-Muskegon. It is the most active entity in the United States in superalloy research. It is also moving more investment overseas, particularly to China, where it has its 5-5-5 program: $5 billion in investment, $5 billion in sales, by 2005.

Pratt & Whitney

Pratt & Whitney has cut back drastically on the number of staff involved in superalloy R&D. It is concentrating on commercial engines, and much of the work of technical personnel in the materials area focuses on solving supplier quality problems.

Rolls-Royce

Formerly GMC's Allison Turbine Division in Indianapolis, Indiana, the now Rolls-Royce-owned facility continues to have its R&D directed largely out of the United Kingdom.

Honeywell

Honeywell—located in Phoenix, Arizona, and formerly known as Garrett Air Research and Allied Signal—did materials research in the past but in recent years cut back and is essentially doing none now. It makes smaller jet engines for helicopters and regional jets and auxiliary power units for aircraft.

Solar Turbines

Solar Turbines, the San Diego, California, producer of land-based turbines, has reduced its alloy development activities dramatically. Today, Solar relies almost exclusively on suppliers for new alloys and parts.

Ladish Corporation

A Milwaukee forger of superalloys, Ladish evaluates new materials and models hot working processes but has only a few materials engineers doing a little R&D.

Wyman-Gordon

A forger of superalloys, Wyman-Gordon evaluates new materials. It is now owned by Precision Castparts, but even before it was purchased, it had essentially eliminated R&D.

OTHER ENTITIES CONDUCTING SUPERALLOY R&D IN THE UNITED STATES

Special Metals Processing Consortium

A consortium of 13 production companies, Special Metals Processing Consortium (SMPC) was formed by the U.S. superalloys industry in 1989 to solve generic preproduction problems in the production of refined superalloy vacuum arc remelt (VAR) and electroslag remelt (ESR) ingots for rotating turbine parts. SMPC's research on fundamental solidification processes during ESR and VAR remelting has helped solve some difficult problems common to all that one company would have found difficult to solve on its own. SMPC has cooperated with the Liquid Metals Processing (LMP) Laboratory at Sandia National Laboratories (SNL).

Over the past 5 years, the number of companies supporting SMPC has dwindled to six or seven through bankruptcy, financial losses, and/or mergers. Government support was obtained from SNL and the Federal Aviation Administration in the past, but SNL intends to close the LMP Laboratory in 2005 and SMPC will no longer have use of the laboratory.

FAA funding for SMPC in the past matched member contributions of $50,000 per member per year; it was made available on a year-to-year basis through specific appropriations in the annual FAA budget. Funding dwindled to $10,000 per member in the early 2000s, and the number of participants decreased as well. Although funding for 2004–2005 was once again at $50,000 per member, in real terms current research support has been cut by more than one-half over the last 15 years. Nevertheless, the research resulted in the best control technology in the world for the production of large VAR ingots, and modeling has eliminated the burden of in-plant experimentation.

Universities

Each of the following American universities boasts a faculty member who has conducted research on some aspect of physical metallurgy or processing of superalloys in recent times: the University of Michigan, Lehigh University, Purdue University, the University of Arizona, the University of Texas, Ohio State University, Pennsylvania State University, the University of Tennessee, Michigan Tech, the University of Pittsburgh, the University of Florida, and Northwestern University.

Government Laboratories

NASA's Glenn Research Center supports university research in superalloys. The Wright-Patterson facility of the Air Force Research Laboratory (AFRL) also supports university research, and the Ohio Aerospace Institute at the University of Dayton conducts testing for AFRL. Oak Ridge National Laboratory is primarily engaged in joining and coating research.

SUPERALLOY RESEARCH AND DEVELOPMENT OUTSIDE THE UNITED STATES

United Kingdom

Virtually all superalloy research in the United Kingdom is funded by Rolls-Royce, which operates a superalloy research laboratory at Cambridge University. Rolls-Royce does not produce alloys but works with primary producers to develop new alloys. It is building a new facility where it will utilize superalloy powders in its next-generation engines. Other research centers in the United Kingdom are at Imperial College, Thermotech Ltd., the University of Southampton, the National Physical Laboratory, the University of Manchester, and Birmingham University.

Germany and Switzerland

Krupp VDM was once a developer of alloys, but it no longer has an R&D operation. Asea Brown Boveri sponsored R&D in the past but has cut back in recent years. Sulzer Metco is doing R&D on coating alloys. Superalloy research continues at the government-supported Max Planck Institute in Aachen, the KFA Laboratory, the Fraunhofer Laboratory, the Technical University of Berlin, the Federal Institute of Materials Research and Testing, the Technical University of Braunschweig, the University of Applied Sciences (Osnabrueck), and the University of Erlangen-Nuremburg.

Italy

Acceria Foroni is a producer of nickel-based superalloys, primarily for the oil and gas industry in the form of casing and tubing, that invested over $200 million in new technology over the last 10 years. It is now investing in large-diameter (33-inch) VAR technology with a view to producing 718 alloy ingots for rotating parts. Acceria Foroni is regarded as a tough competitor in the marketplace.

Japan

Japan has been an active player in superalloy development for many years. Mitsubishi Materials has been and continues to be very active in both the laboratory and the plant. Daido Steel has an interest in superalloy technology and has an ongoing technical exchange with SMC. Hitachi Heavy Industries has eliminated R&D in its laboratory but continues R&D in its plant. Other organizations continuing superalloy R&D include IHI Heavy Industries, the National University for Materials Science, Nagoya University, the Nagoya Institute of Technology, the Tokyo Institute of Technology, and the National Aerospace Laboratory (High Temperature Materials Group).

China

Since the end of the Great Leap Forward, significant research has continued in China on superalloys. The main Chinese superalloy producers include the Fushun Specialty Steel Plant (part of Northeast Specialty Steel Group Company), the Shanghai No. 5 Steel Plant (part of BaoSteel), and the Great Wall Specialty Steel Plant. The main research institutes are the Central Iron and Steel Research Institute, the Beijing Institute of Aeronautic Materials, and the Institute of Metals Research. The main universities for superalloy research are University of Science and Technology Beijing, the Northwestern Polytechnic University, and Northeasten University. There are several plants forging and casting parts for aircraft engines in the different parts of China. The Shanghai No. 5 Steel Plant is a major superalloy production facility with a great deal of the latest and best equipment in the world and a research group said to employ 2,000 professionals. A new pilot plant to produce superalloy castings is being built in Chendu.

Sweden

Demag DeLaval Industrial Machinery and Volvo Aero continue some superalloy research activities.

Korea

Korea is not an important player in the superalloy business, but R&D nonetheless is actively pursued at the Korean Institute of Machinery and Materials.

Canada

Research has been conducted at various universities (the University of British Columbia, the University of Manitoba, Ryerson University, Carleton University), at the National Research Council of Canada's Institute for Aerospace Research, and at BWD Turbines.

Australia

Western Australia Special Alloys (WASA) was started by Pratt & Whitney, Wyman-Gordon, and the government of Western Australia to supply nickel-based alloy VIM-VAR ingots for forging, with the expected advantage of lower cost nickel from Australia. Pratt & Whitney sold its interest to Wyman-Gordon, which then was bought in 2004 by Precision Castparts.

France

DMMP, CEMES/CNRS, the Ecole National Supérieure des Mines, Albert and Duval, LMPM-ENSMA, Institute Laue Langevin, and Snecma all contribute research.

Russia

The Central Boiler and Turbine Institute is active in superalloy research.

PUBLICATIONS

Although not an accurate measure of commercial activity, which is for the most part closely guarded by producers, papers presented at the quadrennial Seven Springs Superalloy Conference, sponsored by The Minerals, Metals and Materials

Society, give some indication of worldwide activity in the field. First held in 1978, the conference has since then attracted a growing audience. A survey of the papers presented indicates that while the number of papers from the United States remained static at about 45 (in the face of a 50 percent decline in U.S. participating companies), the number of papers from overseas has steadily increased. In 2004 there were 66 from outside the country, so that less than 50 percent of the work presented at the conference is of U.S. origin, compared with over 85 percent in 1978.

ISSUES AND CONCLUSIONS

An examination of the current status of U.S. superalloy R&D shows that the workforce decreased significantly between 1994 and 2004—more than 50 percent in most companies and up to 100 percent in some. New or improved materials that are being developed are largely based on improvements to existing alloys or are for the chemical industry, where time to introduction is shorter.

The biggest challenge for superalloy R&D is that it takes a long time to develop a new alloy, on the order of years, and takes even more years to get customer acceptance and large orders. Furthermore, because new alloys often replace existing products, this can mean a very low return on R&D investment. Therefore, faced with bottom-line considerations and the pressure for quarterly profit growth, companies often have little incentive to support R&D activities aimed at the development of new alloys or processes.

An unusually large percentage of the papers at the 2004 Seven Springs Superalloy Conference were joint papers emanating from collaborative research, a sign that fewer companies can support research on their own and are turning to collaboration. While this is better than no research at all, it is symptomatic of the problems facing the industry. Another problem has been a decrease in U.S. government support for superalloy research, leaving companies to pick up the costs of ongoing activities or cut them out altogether. In addition, the time and money required to get a new material approved for aerospace use are a challenge, although DARPA's Accelerated Insertion of Materials program has begun to address these issues.

The lack of new products for the market is both a result of, and a cause of, continued financial weakness in the industry over the past two decades. Future growth, or even maintenance of the current status of the industry, will require patient management, conservative financing, and a stream of new products.

There has been criticism that some international actors are not widely sharing the results of R&D, particularly in the area of process technology, where most of the proprietary knowledge resides.[5] Plugging the flow of information can hinder U.S. companies in their efforts to follow recent developments.

The superalloy industry has evolved to the point that production facilities are now divisions or subsidiaries of larger companies. In the event of an economic slowdown or if their customer base continues to move offshore, companies that are still independent are likely to suffer financially. The use of offsets[6] by superalloy customers—notable for deals involving companies such as GE, Pratt & Whitney, and Boeing—can result in offshore operations being established by U.S. companies in order to win contracts for their own products. These overseas operation could be in direct competition with U.S. operations. Much alloy development is now being done in foreign-government-supported laboratories overseas. Process improvement continues to be done on a limited basis in the United States and is the key to commercial U.S. success for the time being. However, as new alloys are developed elsewhere and U.S. process knowledge diffuses offshore, that competitive edge will disappear. As the manufacturing end of the industry moves overseas there is a risk that the ideas for research, which in the past largely came from working with customers, will move with it. The result may be a drying up of research ideas in American companies.

In conclusion, it is clear that U.S. superalloy R&D declined significantly over the past decade, while Chinese and Japanese superalloy R&D is increasing. U.S. companies that are diversifying into offshore manufacturing will probably survive, and to the extent that they are privy to the knowledge generated in non-U.S. laboratories and plants, they will be able to stay competitive. The foreign superalloy industry is growing and is likely to develop technology that is equivalent to or better than the American technology.

[5]It is noteworthy that no Italians participated in the 2004 Seven Springs conference, yet Acceria Foroni is a major force in the superalloy industry.

[6]Offset obligations are incurred when an agency of a foreign government purchasing a military or commercial product requires that some reciprocal activity take place to help offset the import/export imbalance in their country created by the purchase.

G

Environmental and Safety and Health Regulations

The following is a brief overview, not meant to be all-inclusive, of the U.S. environmental and safety/health regulation issues touched on in Chapter 4 of this report.

ENVIRONMENTAL LAWS

The Clean Air Act (CAA), the Clean Water Act (CWA), the Resource Conservation and Recovery Act (RCRA), and the Comprehensive Environmental Response, Compensation and Liability Act (CERCLA) are the four main environmental laws that affect industry in the United States. Other laws to be aware of include the Emergency Planning and Community Right-to-Know Act (EPCRA), the Superfund Amendments and Reauthorization Act, the Toxic Substances Control Act (TSCA), and the National Environmental Policy Act (NEPA).

Clean Air Act

CAA regulates all sources in the United States that contribute directly or indirectly to the emission of air pollutants. The Environmental Protection Agency (EPA) establishes standards for pollutants believed to adversely affect public health or welfare. States are divided into air quality control regions designated as attainment or nonattainment areas for each criteria pollutant according to whether the air quality in the region is better or worse than the national standard for that pollutant. States are required to have implementation plans and permit programs for controlling certain air emissions, including those from industrial processes such as spray painting, abrasive blasting, and vapor degreasing.

Clean Water Act

CWA regulates all sources of pollution and discharges of pollutants to navigable waters. Any addition of pollution to navigable waters is prohibited, except as provided in the conditions of a National Pollutant Discharge Elimination System permit issued by the EPA or an authorized state agency. The CWA also regulates storm water discharges, dredged or fill material, and oil spills.

Resource Conservation and Recovery Act

RCRA defines which substances are hazardous wastes and regulates their generation, transportation, treatment, storage, and disposal. A hazardous waste is defined as a substance which, because of its quantity, concentration, or other characteristics, may increase mortality or serious illness or pose a hazard to health or the environment. All generators of hazardous waste are responsible for the hazardous wastes they generate from acquisition through disposal. RCRA requires extensive record keeping and reporting.

Comprehensive Environmental Response, Compensation and Liability Act

CERCLA authorizes the President of the United States to clean U.S. facilities at which hazardous substances have been released. The Superfund is used to finance enforcement and cleanup efforts, pending reimbursement by responsible parties. The cost recovery provisions of Superfund create huge potential civil liabilities for business, since they allow the EPA (and others) to sue businesses to recover cleanup costs, which can amount to millions of dollars at a single site. The costs and damages can be recovered from a wide range of persons and businesses associated with the sites, including anyone associated with the contaminated site or the hazardous waste that contaminated the site. The important thing is to learn and review that all the facts before purchasing land or awarding an environmental services contract.

Emergency Planning and Community Right-to-Know Act

EPCRA imposes detailed chemical release and inventory reporting requirements on businesses that handle toxic, hazardous, and extremely hazardous chemicals like hydrofluoric acid (HF). EPCRA establishes reporting or notification requirements for owners or operators of facilities at which certain substances are produced, processed, used, or stored in concentrations above a threshold limit. For example, the concentration threshold for HF is 50 percent. If this limit is exceeded, potential releases must be modeled and the information given to the local governments for the purpose of emergency planning for releases into the environment. The general public must also be notified, in person, if high concentrations of toxic or hazardous substances are being used in their neighborhoods.

Toxic Substances Control Act

TSCA regulates the use of toxic chemicals and releases of chemical substances entering the environment, and it authorizes the EPA to screen all new or imported chemicals entering the marketplace, as well as to evaluate the health and environmental effects of existing chemicals. R&D activities also fall within TSCA's jurisdiction. Manufacturers, and sometimes processors and distributors, have recordkeeping requirements under this law. Polychlorinated biphenyls, lead, asbestos, and many other substances are regulated under this act. Regulated substances are not allowed to be used in new applications. They can continue to be used in current applications, albeit at high cost.

National Environmental Policy Act

NEPA, an environmental law affecting all weapon system programs, regulates federal agencies—for example, DOD and its departments (Air Force, Army, Navy, etc.)—and addresses global problems of pollution prevention, warming, stratospheric ozone depletion, and sustainable growth. A key element of NEPA is that it regulates federal agencies, not the private/public sector. However, industry is becoming more involved with NEPA as a result of the requirements NEPA places on government program managers purchasing industrial products.

NEPA requires that all federally funded programs must conduct an environmental impact analysis prior to all major decisions (Milestones I, II, and III) to consider their impact on safety, human health, and the environment. A federally funded program must obtain a Categorical Exclusion (CATEX), perform an Environmental Assessment (EA), or prepare an Environmental Impact Statement (EIS). A CATEX may be obtained for actions that do not normally have a significant effect individually or cumulatively on the human environment. A CATEX may also be obtained for actions that have been previously found to have no effect based on procedures adopted by DOD for implementing NEPA regulations and that therefore require neither an EA nor an EIS. An EA is a concise public document that provides brief but sufficient evidence and analysis for determining whether to prepare an EIS or a Finding of No Significant Impact; it is one element in achieving compliance with NEPA when no EIS is necessary or it facilitates the preparation of an EIS when one is necessary. When required, an EIS is a major undertaking, often performed by a contractor selected by the program manager specifically for this task.

SAFETY AND HEALTH LAWS

The Occupational Safety and Health Administration (OSHA) requires U.S. employers (1) to furnish employees a place of employment free from recognized hazards that cause or are likely to cause death or serious physical harm and (2) to comply with the occupational safety and health standards promulgated by legislation. In addition, employees are required to comply in their actions and conduct with occupational safety and health standards and all rules, regulations, and orders issued pursuant to relevant legislation. OSHA promulgates standards, rules, and regulations specific to activities and regulates the use of hazardous materials and other materials.

CURRENT ISSUES

A number of current environmental and safety and health issues could have an impact on the materials industry:

- International restrictions on the use of particular materials
  —Restrictions on lead in electronics
  —Restrictions on brominated fire retardants
  —Greenhouse gas restrictions on hydrofluorocarbons.
- OSHA expanded standards and revisions
  —Hexavalent chromium expanded standard
  —Crystalline silica expanded standard
  —Beryllium expanded standard
  —Glycol ether expanded standard
- Water quality
  —Revised spill prevention, control and countermeasures plan rule
  —Impaired waters listing and total maximum daily load (TMDL) development
  —Metal products and machinery effluent limits
  —Effluent limits, construction and development runoff
  —Revisions to local wastewater treatment limits

- Maximum achievable control technology (MACT) rules for air toxics
  - —MACT rule for military surface coatings
  - —MACT rule for boiler and process heaters
  - —MACT rule for engine test cells, new source
  - —MACT rules for urban air toxics
- Emission reductions for ozone and particulate matter
  - —Additional volatile organic compound reductions to meet 8-hour ozone standard
  - —Emission reductions to meet fine particulate standard
- Rules for air toxics residual risk
  - —National Emissions Standards for Hazardous Air Pollutants (NESHAP) residual risk rule for aerospace
  - —NESHAP residual risk rule for chrome plating
  - —NESHAP residual risk rule for halogenated solvent
  - —Whole facility residual risk off-ramp
- Miscellaneous air pollution issues
  - —Phaseout of hydrochlorofluorocarbons
  - —Significant New Alternatives Policy (SNAP) restriction on N-propyl bromide

H

Defining 21st Century Defense Needs

In the summer of 2001, DOD's Defense Science Board (DSB) was asked to recommend how the department's S&T investment should be spent, how much should be invested, and how the military could realize the most value from this investment. The DSB report[1] said that new threats, new adversaries, emerging disruptive technologies, and the speed with which knowledge spreads and technology is applied are among the new challenges to which DOD must respond. The DSB assessed defense and military needs and synthesized nine high-priority military areas:

- Biological warfare defense for immediate detection and defeat.
- Capability to find and correctly identify difficult targets, both static and mobile.
- Support of high-risk operations with systems such as unmanned systems capable of high-risk tactical operations.
- Missile defense that is cost effective and exhibits low leakage against tactical and strategic missiles and unmanned aerial vehicles.
- Affordable precision munitions that are resilient to countermeasures.
- Enhanced human performance that overcomes natural limitations on cognitive ability and endurance.

[1]Defense Science Board, *Defense Science and Technology* (2002). Available at http://www.acq.osd.mil/dsb/reports/sandt.pdf.

- Rapid deployment and employment of forces globally against responsive threats.
- Global effects that can be delivered rapidly, anywhere.

The DSB panel recommended that DOD focus its investment in these nine areas either because they will optimize the payoff or because they are associated with high-risk threats.

Although released in 2002, the DSB report was completed only months before the tragic events of September 11, 2001. While the central assessments of the DSB remain valid, there is no doubt that after those events there was a dramatic refocusing of the nation's attention to national security and, most importantly, homeland security. September 11 caused many new assessments to be undertaken, one of which was a study by the National Research Council of the contributions science and technology might make to counterterrorism.[2]

The aim of the study was to help the federal government—and, more specifically, the Executive Office of the President—to enlist the nation's and the world's scientific and technical community in a timely response to the threat of catastrophic terrorism. The terms of reference for the study called for the preparation of (1) a carefully delineated framework for the application of science and technology for countering terrorism, (2) the preparation of research agendas in nine key areas,[3] and (3) the examination of a series of crosscutting issues. Overall, the authoring committee aimed to identify scientific and technological means by which the nation might reduce its vulnerabilities to catastrophic terrorist acts and mitigate the consequences of such acts when they occur.

The eight panels of preeminent scientists, engineers, and physicians identified 14 "most important technical initiatives":

- Immediate applications of existing technologies
  - —Develop and utilize robust systems for protection, control, and accounting of nuclear weapons and special nuclear materials at their sources.
  - —Ensure production and distribution of known treatments and preventatives for pathogens.

---

[2]*Making the Nation Safer: The Role of Science and Technology in Countering Terrorism*, Committee on Science and Technology for Countering Terrorism, Washington, D.C.: The National Academies Press (2002).

[3]Biological sciences; chemical sciences; nuclear and radiological sciences; information technology and telecommunications; transportation; energy facilities; cities and fixed infrastructure; behavioral, social, and institutional issues; and systems analysis and systems engineering.

—Design, test, and install coherent, layered security systems for all transportation modes, particularly shipping containers and vehicles that contain large quantities of toxic or flammable materials.
—Protect energy distribution services by improving security for supervisory control and data acquisition (SCADA) systems and providing physical protection for key elements of the electric-power grid.
—Reduce the vulnerability and improve the effectiveness of air filtration in ventilation systems.
—Deploy known technologies and standards for allowing emergency responders to reliably communicate with each other.
—Ensure that trusted spokespersons will be able to inform the public promptly and with technical authority whenever the technical aspects of an emergency are dominant in the public's concerns.

- Urgent research opportunities
  —Develop effective treatments and preventatives for known pathogens for which current responses are unavailable and for potential emerging pathogens.
  —Develop, test, and implement an intelligent, adaptive electric-power grid.
  —Advance the practical utility of data fusion and data mining for intelligence analysis, and enhance information security against cyber attacks.
  —Develop new and better technologies (e.g., protective gear, sensors, communications) for emergency responders.
  —Advance engineering design technologies and fire-rating standards for blast- and fire-resistant buildings.
  —Develop sensor and surveillance systems (for a wide range of targets) that create useful information for emergency officials and decision makers.
  —Develop new methods and standards for filtering air against both chemicals and pathogens as well as better methods and standards for decontamination.

Materials research will play a role in most if not all of the 9 high priorities identified by the Defense Science Board and the 14 initiatives identified by the National Research Council study on countering terrorism. It is clear, therefore, that progress in materials research in the United States and abroad will affect the nation's ability to defend itself against emerging threats in the 21st century.

Another report from the National Research Council considered the narrower topic of contributions materials research could make to meet 21st century military needs. DOD asked the National Research Council to conduct a study that would identify and prioritize critical needs for materials and processing R&D to meet 21st century defense needs. The resulting study, released in 2003, identified those needs and explored the revolutionary defense capabilities that could result from R&D in five classes of materials:[4]

- Structural and multifunctional materials,
- Energy and power materials,
- Electronic and photonic materials,
- Functional organic and hybrid materials, and
- Bioderived and bioinspired materials.

In considering the opportunities in these materials subclasses, the study identified the following core tasks for the U.S. military:

- Projecting long-distance military power;
- Maintaining capability to fight far away;
- Coping with the eroding overseas base structure;
- Safeguarding the homeland; and
- Adjusting to major changes in warfare, including joint-service operations, coalition peacekeeping, and an increased number of humanitarian missions.

Furthermore, the study concluded that the following trends in warfare could be expected to continue:

- The need will increase for a precision strike force that can maneuver rapidly and effectively and can survive an attack while far away.
- The force must be able to conceal its activities from an enemy while detecting enemy activities.
- Advances in information technology will increase coordination among forces. Global awareness through real-time networked sensors and communications will facilitate command and control and enable precision strikes.
- Using unmanned vehicles, information will be gathered in new ways, military power will be delivered remotely, and the risk of casualties will be reduced.
- Fighting in urban areas will increase, requiring entirely different strategies and equipment.
- Guerilla warfare will require new strategies and weapons.

---

[4]National Research Council, *Materials Research to Meet 21st Century Defense Needs*, Washington, D.C.: National Academies Press (2003).

The study concluded that DOD needs various types of functionality, alone and in combination, for its military systems. R&D in materials and processes will be required to improve existing materials and achieve breakthroughs in new materials and combinations. Examples of the types of materials needed are as follows:

- Lightweight materials that provide equivalent functionality,
- Materials that enhance protection and survivability,
- Stealth materials,
- Electronic and photonic materials for high-speed communications,
- Sensor and actuator materials,
- High-energy-density materials, and
- Materials that improve propulsion technology.

The drivers of these needs are multiple. Everything from tanks and ships to the equipment worn and carried by a warfighter needs to get lighter without losing functionality. Materials are needed to protect and hide equipment and personnel. The military in the 21st century will need to communicate faster, more reliably, and on a global scale. New threats require new materials for their detection. New tasks will require new weapons and new materials to enable new and better delivery platforms. The new systems of the 21st century military will also need to demonstrate multifuctionality; self-diagnosis and self-healing; low cost and low maintenance; environmental acceptability; and high reliability. The report also concludes that successful research on broad classes of materials and processes will need to be accompanied by the consideration of engineering issues—for example, a new material or process that seems promising in the laboratory would be useless if it could not be manufactured. In addition, the introduction of a new material into a system is more likely to be successful if the new material is integrated into the system or component design effort as early as possible. Finally, end-of-system-life issues must be considered—for instance, recycling or reuse of as much of the system as possible and environmentally conscious disposal of the rest.

More details of this needs-based analysis can be found in the full report, including subpanel reports on the five classes of materials. The report concludes as follows:

> Future defense systems could employ advanced materials that are self-healing, can interact independently with the local environment, and can monitor the health of a structure or component during operation. Advanced materials could act as a host for evolving technologies, such as embedded sensors and integrated antennas. Advanced materials must also deliver traditional high performance in structures; protect against corrosion, fouling, erosion, and fire; control fractures; and serve as fuels, lubricants, and hydraulic fluids. The next 20 years will present the materials community with daunting challenges and opportunities. Requirements for material

producibility, low cost, and ready availability will be much more demanding than they are today. On the other hand, spurred by the accelerated pace of advances in electronics and computation, the performance, life span, and maintainability of materials will be greatly enhanced. Some of the advances will result from R&D undertaken by commercial enterprises for competitive advantage in areas like telecommunications and computation. In other areas, however, DOD may have to bear the funding burden directly. In these special areas, considerable funding will be necessary not only to identify critical new materials, but also to accelerate their progress through development to applications in the defense systems of the future.